Further volumes in this series:

Volume 1
A. Avižienis, H. Kopetz, J. C. Laprie (eds.)
The Evolution of Fault-Tolerant Computing

Volume 2
U. Voges (ed.)
Software Diversity in Computerized
Control Systems

Volume 3
P. A. Lee, T. Anderson
Fault Tolerance - Principles and Practice
Second, revised edition

Volume 4
A. Avižienis, J. C. Laprie (eds.)
Dependable Computing for Critical Applications

Dependable Computing and Fault-Tolerant Systems

Edited by
A. Avižienis, H. Kopetz, J. C. Laprie

Volume 5

Springer-Verlag Wien GmbH

J. C. Laprie (ed.)

Dependability: Basic Concepts and Terminology

in English, French, German, Italian and Japanese

Springer-Verlag Wien GmbH

Dr. Jean-Claude Laprie, LAAS-CNRS, Toulouse, France

With 3 Figures

Originally published by Springer Verlag Vienna 1992
Softcover reprint of the hardcover 1st edition 1992

Printed on acid-free paper

ISSN 0932-5581
ISBN 978-3-7091-9172-9 ISBN 978-3-7091-9170-5 (eBook)
DOI 10.1007/978-3-7091-9170-5

IFIP WG 10.4
Dependable Computing and Fault Tolerance

J.C. Laprie (ed.)

Dependability: Basic Concepts and Terminology

Sûreté de Fonctionnement : Concepts de Base et Terminologie

Zuverlässigkeit: Grundkonzepte und Terminologie

Garanzia di Funzionamento: Concetti Base e Terminologia

デペンダビリティ： 概念と関連用語

Contributors

T. Anderson (The University of Newcastle upon Tyne, UK)
A. Avizienis (UCLA, Los Angeles, California, USA)
W.C. Carter (Bailey Island, Maine, USA)
A. Costes (LAAS-CNRS, Toulouse, France)
F. Cristian (IBM Research, San Jose, California, USA)
Y. Koga (National Defense Academy, Yokusuka, Japan)
H. Kopetz (Technische Universität Wien, Austria)
J.H. Lala (C.S. Draper Lab., Cambridge, Massachusetts, USA)
J.C. Laprie (LAAS-CNRS, Toulouse, France)
J.F. Meyer (University of Michigan, Ann Arbor, Michigan, USA)
B. Randell (The University of Newcastle upon Tyne, UK)
A.S. Robinson (STDC, Reston, Virginia, USA)
L. Simoncini (University of Pisa, Italy)
U. Voges (KFK, Karlsruhe, Germany)

FOREWORD

This volume gathers a text resulting from several years of work within the Working Group 10.4 of IFIP, *Dependable Computing and Fault Tolerance*, together with its French, German, Italian, and Japanese versions. The aim is to give precise definitions characterizing the dependability of computing systems. Dependability is first introduced as a global concept and a set of basic definitions is given. Those definitions are then commented, and supplemented, in the subsequent sections dealing respectively with the impairments to dependability (faults, errors, failures), the means for dependability (fault-prevention, fault-tolerance, fault-removal, fault-forecasting), and the attributes of dependability (reliability, availability, safety, security). The 116 definitions given throughout the text are recapitulated in glossaries, and a five-language cross-index is provided.

What is presented would not exist without the many discussions held with many colleagues. The prominent role of the contributors listed on the title page is acknowledged, through either their written contributions and comments, or their help in developping ideas during discussions. It is a pleasure to extend these acknowledgements to all members of WG 10.4, especially to David Morgan for his seeding role, and to the members of the research group "Dependable Computing and Fault Tolerance" of LAAS, of which Jean Arlat, Christian Beounes, Yves Deswarte, Jean-Charles Fabre, David Powell, and Pascale Thevenod deserve a special attention. Of particular importance is the work performed by my colleagues who kindly accepted to produce the versions gathered here: Yoshiaki Koga produced the Japanese version, Luca Simoncini produced the Italian version, Udo Voges produced the German version with the help of Winfried Görke and Hermann Kopetz. I was greatly helped by Alain Costes in producing the French version, and his endless readiness to discuss this topic over the years deserve a special attention. Particular thanks go also to Joëlle Penavayre for her help in the formatting task.

Toulouse, July 1991

Jean-Claude Laprie

Contents

Garanzia di Funzionamento: Concetti Base e Terminologia 153

DEPENDABILITY:

BASIC CONCEPTS

AND TERMINOLOGY

INTRODUCTION

This document is aimed at giving informal but precise definitions characterizing the various attributes of computing systems dependability. It is a contribution to the work undertaken within the "Reliable and Fault Tolerant Computing" scientific and technical community [Avi 67, Jes 77, Mel 77, Avi 78, Ran 78, Car 79, And 81, FTC 82, Sie 82, Cri 85a, Lap 85, Avi 86, Lap 89] in order to propose clear and widely acceptable definitions for some basic concepts.

Dependability is first introduced as a global concept which subsumes the usual attributes of reliability, availability, safety, security. The basic definitions given in the first section are then commented, and supplemented by additional definitions, in the subsequent sections. A glossary is given in the annex, which recapitulates the definitions given throughout the document. The presentation has been structured in order to avoid forward referencing. Boldface characters are used when a term is defined, italic characters being an invitation to focus the reader's attention. The guidelines which have governed this presentation can be summed up as follows:

- search for a reduced number of concepts enabling the dependability attributes to be expressed;
- use of terms which are identical to — whenever possible — or as close as possible to those generally used; as a rule, a term which has not been defined retains its ordinary sense (as given by any dictionary);
- emphasis on integration [Gol 82, Ran 86] (as opposed to specialization) through the independence of the given definitions with respect to the classes of faults.

This document can be seen as a minimum consensus within the community in order to facilitate fruitful interactions; in addition this document is hoped to be suitable a) for use by other bodies (including standard organizations), and b) for educational purposes. In this view, the associated terminology effort is not an end in itself: words are only of interest in so far as they unequivocally label concepts, and enable ideas and viewpoints to be shared. The document is devoid of any pretension of being a state-of-the-art or "Tablets of Stone": the concepts that are presented have to evolve with technology, and with our progress in understanding and mastering the specification, design and assessment of dependable computer systems.

What is exposed in this document would not exist without the many discussions held with many colleagues, especially the members of IFIP WG 10.4.

1- BASIC DEFINITIONS

Dependability is defined as the trustworthiness of a computer system such that reliance can justifiably be placed on the service it delivers [Car 82]. The **service** delivered by a system is its behavior *as it is perceived* by its user(s); a **user** is another system (human or physical) which *interacts* with the former.

Depending on the application(s) intended for the system, different emphasis may be put on different facets of dependability, i.e. dependability may be viewed according to different, but complementary, *properties,* which enable the *attributes* of dependability to be defined:

- with respect to the *readiness for usage*, dependable means **available**;
- with respect to the *continuity of service,* dependable means **reliable**;
- with respect to the *avoidance of catastrophic consequences on the environment,* dependable means **safe**;
- with respect to the *prevention of unauthorized access and/or handling of information*, dependable means **secure**.

A system **failure** occurs when the delivered service no longer complies with the **specification**, the latter being an agreed description of the system's expected function and/or service. An **error** is that part of the system state which is liable to lead to subsequent failure: an error affecting the service is an indication that a failure occurs or has occurred. The adjudged or hypothesized cause of an error is a **fault**.

The development of a dependable computing system calls for the *combined* utilization of a set of methods which can be classed into:

- **fault prevention**: how to prevent fault occurrence or introduction;
- **fault tolerance**: how to provide a service complying with the specification in spite of faults;
- **fault removal**: how to reduce the presence (number, seriousness) of faults;
- **fault forecasting**: how to estimate the present number, the future incidence, and the consequences of faults.

Fault prevention and fault tolerance may be seen as constituting dependability **procurement**: how to *provide* the system with the ability to deliver a service complying with the specification; fault removal and fault forecasting may be seen as constituting dependability **validation**: how to *reach confidence* in the system's ability to deliver a service complying with the specification.

Reliance on a system's service, and justification for reliance, are (or at any rate should be) based on the *assessment* of the system, conducted primarily with respect to the *attributes* of dependability which are pertinent to the system's service.

The notions introduced up to now can be grouped into three classes (figure 1):

- the **impairments** to dependability: faults, errors, failures; they are undesired — but not in principle unexpected — circumstances causing or resulting from un-dependability (whose definition is very simply derived from the definition of dependability: reliance cannot, or will not any longer, be placed on the service);
- the **means** for dependability: fault prevention, fault tolerance, fault removal, fault forecasting; these are the methods and techniques enabling one a) to provide the ability to deliver a service on which reliance can be placed, and b) to reach confidence in this ability.
- the **attributes** of dependability: availability, reliability, safety, security; these a) enable the properties which are expected from the system to be expressed, and b) allow the system quality resulting from the impairments and the means opposing to them to be assessed.

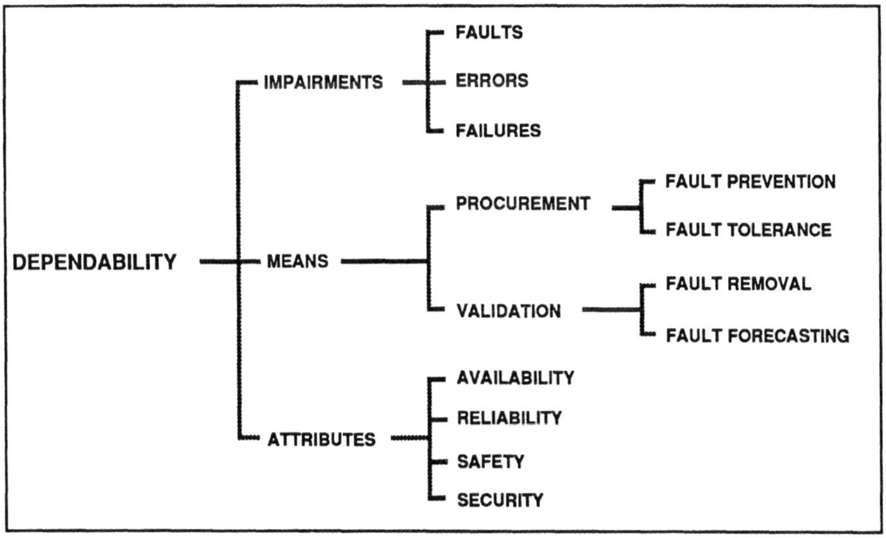

Figure 1 - The dependability tree

2- On The Introduction of Dependability as a Generic Concept

A natural tendency of any emerging scientific or technical discipline is, in a first step, to restrict its field of investigation in order to make — rapid — progress in solving the associated problems. Then comes a time when its interactions with other disciplines can no longer be ignored. A great temptation is then to declare those other disciplines as being "special cases" of the particular discipline. This usually results in large debates, often conducted by the adherents of each discipline each in their own jargon. This was the case for reliability, availability, safety, and security of computing systems. Initially, the main concern was to have computing systems work: *reliability*. As computing systems became reliable, their services were used and needed on a regular basis, so *availability* became essential. The utilization of computing systems in critical applications brought in the concern for *safety*. The safest system is often the one which does not do anything, which is not very helpful; so, people concerned with safety tend to consider reliability a subset of safety. The advent of distributed systems has exacerbated the *security* issues. Again, a secure system is intended to fulfill functions; in addition, security violations can be catastrophic; so, people concerned with security tend to consider safety and reliability as subsets of security.

However, the relations between reliability, safety and security are more complex than a simple dependence. Let us consider the example of the so-called "softbombs", i.e. faults deliberately introduced in a computing system in order to provoke, at a moment chosen by the "terrorist" — and under his/her control — a system failure, of consequences preferably felt by the user as non-catastrophic (until he becomes aware of the failure causes). This example clearly involves reliability, safety, and security, in an intricate and varying manner depending upon the viewpoint considered[1]. What is certain is that the user cannot, or should not, place reliance on the service delivered by such a system, which is not *dependable,* in the primary sense of the word.

The preceding discussion clearly shows that it is *not* this document's intention to contribute to the controversy concerning whether reliability is a broader concept than safety or vice-versa, and similarly when security is added. What is essential in the relationship of reliability, safety and security and the notion of dependability is that the former are *attributes* of the latter: as

1 It is also noteworthy that the events reported in the section on "Risk to the Public in Computer Systems" of the ACM Software Engineering Notes relate to reliability, to safety, and to security.

such, it is hoped that cross-fertilization will be favored. This is the main reason for the addition of another word to an already long list — reliability, availability, safety, security, etc. An additional reason for the introduction of dependability as a generic concept is the willingness to relieve reliability from its global, etymological, meaning — ability to rely upon — in order to concentrate on its widely (and historically recent) accepted relation to continuity of service, with respect to both its general understanding — reliable system — and its probabilistic definition — system reliability[2].

Although "dependability" is synonymous to reliability, it has a connotation of dependence. This may be felt as a negative connotation at first sight, when compared to the positive notion of trust as expressed by reliability, but it does highlight our society's ever increasing dependence upon sophisticated systems in general and especially upon computing systems. Continuing with etymological considerations, "to rely" comes from the French "relier", itself from the Latin "religare", to bind back: re-, back and ligare, to fasten, to tie. The French word for reliability, "fiabilité", can be traced back to the 12th century, to the word "fiableté" whose meaning was "character of being trustworthy"; the Latin origin is "fidare", a popular verb meaning "to trust". In the light of these etymological considerations, it can only be regretted that the definition of reliability currently employed in many engineering fields[3] has substituted the notion of "ability" for the notion of "trust", for (at least) the two following reasons:

a) from the viewpoint of the system user, what is of real interest is not so much the *ability* to provide functionalities, as the *service* which is *actually* delivered to the user;
b) from the viewpoint of the system producer who is willing to admit the possible existence of faults in his design, the interpretation of the term "ability" must be questionable, despite the fact that it has been adopted in software engineering glossaries (see e.g. [IEE 82]).

---

2 It is interesting to note that:
a) most books having the word "reliability" in their title actually deal with how to evaluate, measure, predict the reliability of systems, not really with how to build reliable systems;
b) viewing dependability as a more general concept than reliability, availability, etc. and embodying the latter terms, has already been attempted in the past (see e.g. [Hos 60]); this was however attempted with less generality than here, since the goal was to define a measure embodying availability and reliability, and security was not of concern.

3 For example, «Reliability: The ability of an item to perform a required function under given conditions for a given time interval» [IEC 85].

3- On System Function, Behavior, Structure and Specification

Up to now, a **system** has been — implicitly — considered as a whole, emphasizing its externally perceived behavior. A definition complying with this "black box" view is: an entity having interacted or interfered, interacting or interfering, or likely to interact or interfere with other entities, i.e., with other systems. These other systems have been, are, or will constitute the **environment** of the considered system[4]. A system **user** is that part of the environment which *interacts* with the considered system: the user provides inputs to and/or receives outputs from the system, its distinguishing feature being to *use the service* delivered by the system.

The **function** of a system is what the system *is intended for* [Kui 85]. The **behavior** of a system is what the system *does*. What *makes it do what it does* is the **structure** of the system [Zie 76]. Adopting the spirit of [And 81], a system, from a structural ("white box" or "glassbox") viewpoint, is a set of components bound together in order to interact; a **component** is another system, etc. The recursion stops when a system is considered as being **atomic**: any further internal structure cannot be discerned, or is not of interest and can be ignored. The term "component" has to be understood in a broad sense: layers of a system as well as intralayer components; in addition, a component being itself a system, it embodies the interrelation(s) of the components of which it is composed. A more classical definition of system structure is what a system *is*. Such a definition fits in perfectly when representing a system without accounting explicitly for any impairments to dependability, and thus in the case where the structure is considered as *fixed*. We do not want to restrict ourselves to systems whose structure is fixed. In particular, we need to allow for structural changes caused by, or resulting from, dependability impairments. It thus appears that a structure may have states[5]. Hence a definition for the notion

---

4 a) Giving recursive definitions is not for recursion's sake. The aim is to emphasize relativity with respect to the adopted viewpoint. So is it for the notion of system: a given system's boundaries may vary depending on whether it is viewed by its designer(s), by its user(s), by its maintenance crew, etc.

b) The past, present and future forms are employed in order to stress that a system's environment may vary with time, especially with respect to the phases of its life-cycle. For instance, the notion of "programming environment" fits into the given definition, as well as the physical environment a system is confronted with during operation.

5 a) It could therefore be said that a "structure" has also a "behavior", especially with respect to the dependability impairments, even if the considered velocities of evolution with respect i) to the user's request on one hand, and ii) to the impairments on the other, are —hopefully— different.

b) The given definition enables other types of systems with varying structures to be embodied, e.g. adaptive —especially knowledge-based— systems.

of **state**: a condition of being *with respect to a set of circumstances*, whether of behavior or of structure[6].

From its very definition (the user-perceived behavior), the service delivered by a system is clearly an *abstraction* of the system's behavior. It is noteworthy that this abstraction is highly dependent on the application that the computer system supports. An example of this dependence is the important role played in this abstraction by time: the time granularities of the system and of its user(s) are generally different, and the difference varies from one application to another. In addition, the service is of course not restricted to the outputs only, but encompasses all interactions which are of interest to the user; for instance, scanning sensors clearly is part of the service expected from a monitoring system.

We have up to now used the singular for function and service. A system generally fulfils more than one function, and delivers more than one service. The function and the service can be thus seen as composed of function items and of service items. For the sake of simplicity, we shall simply use the plural — functions, services — when it is worth distinguishing several items of function or of service.

Of special interest with respect to dependability are timeliness properties. A **real-time function or service** is a function or service that is required to be fulfilled or delivered within finite time intervals *dictated by the environment*, and a **real-time system** is a system which fulfils at least one real-time function or delivers at least one real-time service [PDC 90].

The *specification* of a system describes the system's expectations in terms of a) either or both its expected function and its expected service, and of b) conditions in — or under — which they have to be fulfilled or delivered: environment, exposure time, performance, observability, etc. The function and/or the service are usually first specified in terms of what *should* be fulfilled or delivered regarding the system's primary aim(s). When considering safety- or security-related systems, this specification is generally completed with what *should not* happen (e.g. the hazardous states from which a catastrophe may ensue, or the disclosure of sensitive information). Such a specification may in turn lead to specifying — additional — functions or services that the system *should* fulfill or deliver in order to reduce the likelihood of what

6 This definition is aimed at emphasizing the relativity of the notion of state, which depends directly upon the phenomena and circumstances considered; e.g. state wrt to computation activity, state wrt to failure occurrence.

should not happen (e.g. checking the authorization rights of a user and authenticating him/her).

In addition, these various specifications may be:

a) expressed according to various degrees of detail : requirement specification, design specification, realization specification, etc.
b) decomposed according to the absence or the presence of failure; the former case relates to what is usually termed the *nominal* mode of operation, and the latter case may relate to the *degraded* mode of operation if the surviving resources are not any longer sufficient for delivering the nominal service(s).

As a consequence, there is usually not a single specification, but several, and, clearly, a system may fail with respect to some of these multiple specifications, and still comply with the others.

It is essential that a specification be *agreed upon* by two persons or corporate bodies — in fact, legal personnae: the system supplier (in a broad sense of the term: designer, builder, vendor, etc.) and its human user(s)[7]. The agreement is necessary in order that the specification can serve as a basis for adjudicating whether the delivered service is acceptable or not, or, equivalently, whether a failure has occurred or not. What can be judged as an acceptable service with respect to a specification at a given level of detail may not comply with the specification at a less detailed level, because of mistakes occurred when detailing the specification, resulting in fact in *specification faults*; specification faults may in turn affect any of the various specifications. More generally, a specification cannot be claimed to be immutable once established. This would be simple ignorance of the facts of life, which imply *change*. The changes may be motivated by modifying the system requirements: modification of the expected function and/or service, or correction of some faults[8]. Once more, what is important is that the specification is (again) agreed upon.

Based on the preceding view of system structure, the notions of function, of service and of their specification apply equally naturally to the components. This is especially interesting in the design process, when off-the-shelf components, either hardware or software are used: what is more of interest to the

7 The agreement may be implicit, as when purchasing a system which comes with its specification and user's manual, or when using off-the-shelf systems.

8 We are thus faced with a circular problem: a reference is needed for adjudicating whether a delivered service is acceptable or not, and this reference may be faulty. Improving specifications has long been devoted a significant amount of attention, including proposals for life-cycle models aimed at this objective [Boe 88].

designer is the function and/or the service they are able to provide, rather than their detailed (internal) behavior.

4- THE IMPAIRMENTS TO DEPENDABILITY

4.1 - Faults

Faults and their sources are extremely diverse. They can be classified according to three main viewpoints which are their nature, their origin and their persistence.

The *nature* of faults leads one to distinguish:

- **accidental faults**, which appear or are created fortuitously;
- **intentional faults**, which are created deliberately, presumably malevolently.

The *origin* of faults may itself be decomposed into three viewpoints:

- the *phenomenological causes*, which lead one to distinguish [Avi 78]:
  - **physical faults**, which are due to adverse physical phenomena,
  - **human-made faults**, which result from human imperfections;
- the *system boundaries*, which lead one to distinguish:
  - **internal faults**, which are those parts of the state of a system which, when invoked by the computation activity, will produce an error,
  - **external faults**, which result from interference or from interaction with its physical (electromagnetic perturbations, radiation, temperature, vibration, etc.) or human environment;
- the *phase of creation* with respect to the system's life, which leads one to distinguish:
  - **design faults**, which result from imperfections arising either a) during the development of the system (from requirement specification to implementation) or during subsequent modifications, or b) during the establishment of the procedures for operating or maintaining the system;
  - **operational faults**, which appear during the system's exploitation.

A distinction can also be made with respect to temporal *persistence* of faults, leading to:

- **permanent faults**, whose *presence* is not related to pointwise conditions whether they be internal (computation activity) or external (environment),
- **temporary faults**, whose presence is related to such conditions, and are thus present for a limited amount of time.

Security issues are dominated by — but not restricted to — intentional faults, which are clearly *human-made* faults. Intentional faults can be either internal or external; typical examples are:

- concerning internal faults, the incorporation of **malicious logic** (e.g. the so-called "Trojan horses"), which is an intentional *design* fault;
- concerning external faults, an **intrusion** which is an intentional *operational external* fault.

In order to be "successful", intentional faults may take advantage of accidental faults, e.g. an intrusion exploiting a security breach due to an accidental design fault; there are interesting and obvious similarities between this example and an accidental temporary external fault "exploiting" a lack of shielding.

It could be argued that introducing the phenomenological causes in the classification criteria of faults may lead recursively "a long way back", e.g. why do programmers make mistakes? why do integrated circuits fail? The very notion of fault is *arbitrary,* and is in fact a facility provided for stopping the recursion. Hence the definition given: *adjudged or hypothesized* cause of an error. This cause may vary depending upon the chosen viewpoint: fault tolerance mechanisms, maintenance engineers, repair shop, developer, semiconductor physicist, etc. In our view, recursion stops at *the cause which is intended to be prevented or tolerated.* This view provides consistency with the distinction between human-made and physical faults: a computing system is a human artifact and as such any fault in it or affecting it is ultimately human-made since it represents human inability to master all the phenomena which govern the behavior of a system. In an absolute sense, a distinction between physical faults and human-made faults (especially design faults) may be considered unnecessary; however, this distinction is of importance when considering the (current) methods and techniques for procuring and validating dependability. If the recursion mentioned above is not stopped, then *a fault is nothing else than the consequence of a failure of some other system* (including the developer) *that has delivered or is now delivering a service to the given system.*

Examples of the preceding discussion follow:

- a design fault results from a designer failure;
- a physical internal fault is due to a hardware component failure, which is itself the consequence of (an) error(s) at the electrical or electronic level (the "physics reliability" community rarely characterizes failures as "sudden and nonpredictable"), in turn originating from physico-chemical disorders, again originating from the hardware production, or from — the limits of — our knowledge of semiconductor physics; physical internal faults can be seen as *recurrent* faults as long as their repair consists of replacing the failed component by an identical — non-failed — component: the new component may fail again similarly in the future unless the failure cause has been traced in the design and production process and removed, thus leading to a modified component with a lower failure likelihood;
- a physical or human-made external fault is in fact a design fault: the inability to foresee all the situations the system will be faced with during its operational life, or the refusal to consider some of them (e.g. for economic reasons); for instance:
  - in the case of an electromagnetic perturbation: is it an external fault or a design fault, i.e. the lack of adequate shielding?
  - in the case of a failure caused by an operator typing a single inappropriate character: is it an interaction fault or a design fault, i.e. the lack of confirmation asked by the system [Nor 83]?

<u>The temporal persistence viewpoint</u> deserves the following comments:

1) Temporary external faults originating from the physical environment are often termed **transient faults.**
2) Temporary internal faults are often termed **intermittent faults**; such faults result from the presence of rarely occurring combinations of conditions; examples are a) "pattern sensitive" faults in semiconductor memories, changes in the parameters of a hardware component (effects of temperature variation, delay in timing due to parasitic capacitance, etc.), or b) situations — affecting either hardware or software — occurring when the system load goes beyond a certain level, such as marginal timing and synchronisation. The very notion of intermittent faults is in an absolute sense arbitrary: such faults are nothing other than — permanent — faults whose conditions of activation cannot be reproduced or which occur rarely enough; however, as already pointed out for the distinction between physical faults and design faults, their consideration is a useful facility.

From the above discussions, it appears that *any fault can be viewed as a permanent design fault.* This is indeed true in an absolute sense, but is not very helpful for the developers and assessors of a system.

Figure 2 summarizes the various classes of faults which have been dealt with, with respect to the various viewpoints which have been considered. If all the combinations of fault classes according to the 5 viewpoints of figure 2 were possible, there would be 32 different fault classes. In fact, the number of likely combinations is more restricted: 11 combinations are indicated by the rows of figure 3, which also gives the usual labelling of these combinations — *not their definition.* These labels are commonly used in order to express in a condensed manner the result of combining several viewpoints.

4.2 - Errors

An error was defined as being *liable* to lead to subsequent failure. Whether or not an error will actually lead to a failure depends on three major factors:

1) The system composition, and especially the nature of the existing redundancy:
   - *intentional* redundancy (introduced to provide fault tolerance) which is explicitly intended to prevent an error from leading to failure,
   - *unintentional* redundancy (it is practically difficult if not impossible to build a system without any form of redundancy[9]) which may have the same — unexpected — result as intentional redundancy.
2) The system activity: an error may be overwritten before creating damage.
3) The definition of a failure from the user's viewpoint: what is a failure for a given user may be a bearable nuisance for another one. Examples are a) accounting for the user's time granularity: an error which "passes through" the system-user(s) interface may or may not be viewed as a failure depending on the user's time granularity, b) the notion of "acceptable error rate" — implicitly before considering that a failure has occurred — in data transmission. This discussion explains why it is often desirable to explicitly mention in the specification such conditions as the maximum outage time (related to the user time granularity).

---

9 A classical problem in hardware testing is the removal of such "false redundancies", whose effect may be to mask faults, and as such to make the task of test pattern generation more complicated.

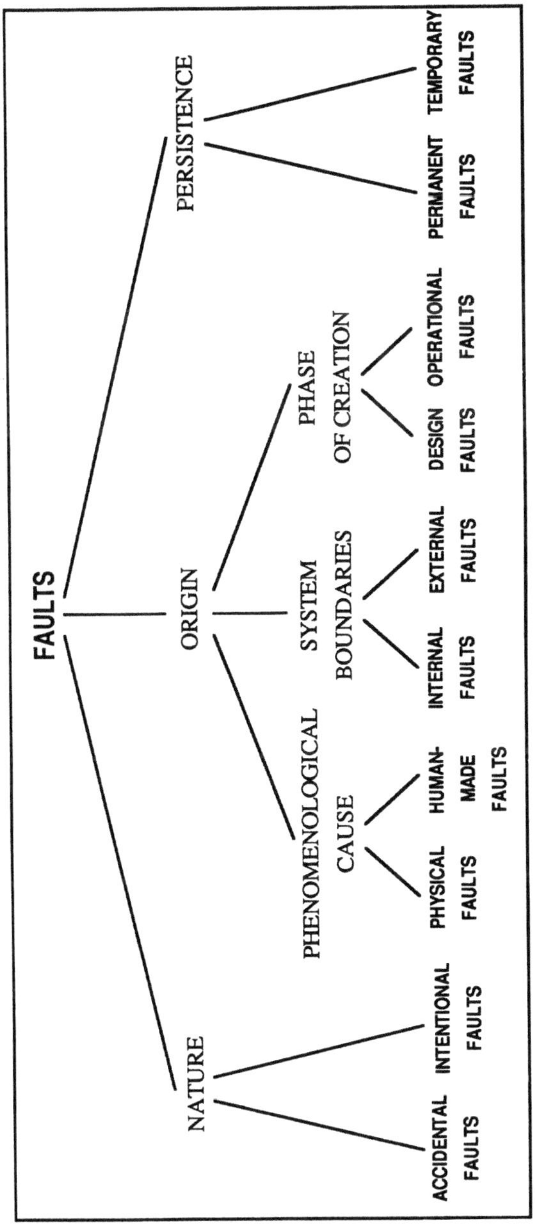

<u>Figure 2</u> - The classes of faults according to various viewpoints

Nature		Origin						Persistence		Usual Labelling
		Phenomenological cause		System Boundaries		Phase of creation				
Accidental Faults	Intentional Faults	Physical Faults	Human-made Faults	Internal Faults	External Faults	Design Faults	Operational Faults	Permanent Faults	Temporary Faults	
✓		✓		✓			✓	✓		Physical faults
✓		✓			✓		✓	✓		Physical faults
✓		✓			✓		✓		✓	Transient faults
✓		✓		✓			✓		✓	Intermittent faults
✓			✓	✓		✓			✓	Intermittent faults
✓			✓	✓		✓		✓		Design faults
✓			✓		✓		✓		✓	Interaction faults
	✓		✓	✓		✓		✓		Malicious Logic
	✓		✓	✓		✓			✓	Malicious Logic
	✓		✓		✓		✓	✓		Intrusions
	✓		✓		✓		✓		✓	Intrusions

Figure 3 - The classes of faults resulting from the combinations according to the various viewpoints of figure 2

4.3 - Failures

Based on the given definition of a system's structure, the discussion of whether "failure" applies to a system or a component is simply irrelevant, since a component is itself a system. When atomic systems are dealt with, the notion of an "elementary" failure comes naturally.

A system may not, and generally does not, always fail in the same way. The ways a system can fail are its *failure modes*, which may be characterized according to three viewpoints: domain, perception by the system users, and consequences on the environment.

The *failure domain* viewpoint leads one to distinguish:

- **value failures**: the value of the delivered service does not comply with the specification;
- **timing failures**: the timing of the service delivery does not comply with the specification.

Such general definitions (non compliance with the specification) apply to **arbitrary failures**. Refined modes of failures can be distinguished. For instance, the notion of timing failure may be refined into *early* timing failure or *late* timing failure, depending on whether the service is delivered too early or too late. A class of failures relating to both value and timing are the **stopping failures**: system activity, if any, is no longer perceptible to the users, and a constant value service is delivered; the constant value delivered may vary according to the application, e.g. last correct value, some predetermined value, etc. A special case of stopping failures is constituted by **omission failures** [Cri 85b, Ezh 86]: no service is delivered. Such a failure can be seen as a common limiting case for both value failures (null value) and timing failures (infinitely late failure); a *persistent* omission failure is a **crash failure**. A system whose failures can be — or more generally are to an acceptable extent — only stopping failures, is a **fail-stop system**, and a system whose failures can be — are to an acceptable extent — only crash failures, is a **fail-silent system**[10] [Pow 88].

When a system has several users, the *failure perception* viewpoint leads one to distinguish:

- **consistent failures**: all system users have the same perception of the failures;

[10] The notion of fail-stop *processor* as defined in [Sch 83] in the context of distributed systems can be seen as an example of a fail-silent system.

- **inconsistent failures**: the system users may have different perceptions of a given failure; inconsistent failures are usually termed, after [Lam 82], *Byzantine failures.*

The failure **severities** result from grading the *consequences of the failures* upon the system environment. They thus enable the failure modes to be ordered. A special case of great interest is that of systems whose failure modes can be grouped into two classes whose severities differ considerably:

- **benign failures**, where the consequences are of the same order of magnitude (generally in terms of cost) as the benefit provided by service delivery in the absence of failure;
- **catastrophic failures**, where the consequences are incommensurably greater than the benefit provided by service delivery in the absence of failure.

A system whose failures can only be — or more generally are to an acceptable extent — benign failures is a **fail-safe system**. The notion of failure severity enables the notion of criticality to be defined: the **criticality** of a system is the highest severity of its (possible) failure modes[11]. The relation between failure modes and failure severities is highly application-dependent. However, there exist a broad class of applications where inoperation is considered as being a naturally safe position (e.g. ground transportation, energy production), whence the direct correspondence which is often made between fail-stop and fail-safe [Min 67, Nic 89].

4.4- Fault pathology

The creation and manifestation mechanisms of faults, errors, and failures may be summarized as follows:

1) A fault is **active** when it produces an error. An active fault is either a) an internal fault which was previously **dormant** and which has been activated by the computation process, or b) an external fault. Most internal faults cycle between their dormant and active states. Physical faults can directly affect the hardware components only, whereas human-made faults may affect any component.

---

[11] As an example, the criticality levels accepted by the aviation community are defined as follows [RTC 85]:

- <u>critical</u>: functions for which the occurrence of any failure would prevent the continued safe flight and landing of the aircraft;
- <u>essential</u>: functions for which the occurrence of any failure would reduce the capability of the aircraft or the ability of the crew to cope with adverse operating conditions;
- <u>non essential</u>: functions for which failure could not significantly degrade aircraft capability or crew ability.

2) An error may be latent or detected. An error is **latent** when it has not been recognized as such; an error is **detected** by a detection algorithm or mechanism. An error may disappear before being detected. An error may, and in general does, propagate; by propagating, an error creates other — new — error(s). During operation, the presence of active faults is determined only by the detection of errors.
3) A failure occurs when an error "passes through" the system-user interface and affects the service delivered by the system. A component failure results in a fault a) for the system which contains the component, and b) as viewed by the other component(s) with which it interacts; the failure modes of the failed component then become fault types for the components interacting with it.

These mechanisms enable the "fundamental chain" to be completed:

$$\cdots \rightarrow \text{failure} \rightarrow \text{fault} \rightarrow \text{error} \rightarrow \text{failure} \rightarrow \text{fault} \rightarrow \cdots$$

Some illustrative examples of fault pathology:

- the result of a programmer's *error* is a *(dormant) fault* in the written software (faulty instruction(s) or data); upon activation (invoking the component where the fault resides and triggering the faulty instruction, instruction sequence or data by an appropriate input pattern) the fault becomes *active* and produces an error; if and when the erroneous data affect the delivered service (in value and/or in the timing of their delivery), a *failure* occurs;
- a short circuit occurring in an integrated circuit is a *failure* (with respect to the specification of the circuit); the consequence (connection stuck at a Boolean value, modification of the circuit function, etc.) is a *fault* which will remain dormant as long as it is not activated, the continuation of the process being identical to the previous example;
- an electromagnetic perturbation of sufficient energy is a *fault;* this fault may
  a) directly create an *error*, e.g. by electromagnetic interference with the electrical charges circulating along wires,
  b) create another, (internal) fault; for instance, if the perturbation acts on a memory's inputs in the write position in changing some bit values, these errors will subsequently remain as faults in the memory; the latter faults will remain dormant until the particular memory location(s) are read; the error-failure sequence from the external transient fault to the internal fault still exists, at the electronic level;

- an inappropriate man-machine interaction performed by an operator during the operation of the system is a *fault* (from the system viewpoint); the resulting altered processed data is an *error;* etc.
- a maintenance or operating manual writer's *error* may result in a *fault* in the corresponding manual (faulty directives) which will remain dormant as long as the directives are not acted upon in order to deal with a given situation, etc.

From the above examples, it is easily understood that the fault dormancy may vary considerably, depending upon the fault, the given system's utilization, etc.

The man-made faults can be either accidental or intentional. The previous example relating to a programmer's error and its consequences may be rephrased as follows: a logic bomb is created by a malicious programmer; it will remain dormant until activated (e.g. at some predetermined date); it then produces an error which may lead to a storage overflow or to slowing down the program execution; as a consequence, service delivery will suffer from a so-called denial-of-service, a special type of failure.

These examples were deliberately kept simple. Real life is, as usual, much more complicated; four examples:

a) a given fault in a given component may result from different possible sources; for instance, a permanent fault in a physical component — e.g. stuck at ground voltage — may result from:
   - a physical failure (e.g. caused by a threshold change),
   - an error caused by a design fault — e.g. faulty microinstruction — propagating "top-down" through the layers and causing a short between two circuit outputs for a duration long enough to provoke a short-circuit having the same consequence as the threshold change;
b) a fault of a given class may, through error propagation, create a fault of another class; for instance, the fault having led to an error during the execution of the microinstruction in the preceding example could have been a transient fault;
c) some viewpoints may become — temporarily at least — of lesser importance during the propagation process; for instance, when dealing with external faults producing input errors during execution of a software component (thus, invoking it in its so-called exceptional input domain [Cri 80]), the fact that the fault is physical or human-made may not be of importance for the failure behavior of the given component;

d) a failure often results from the combined action of several faults; this is especially true when considering security issues: a trap-door (i.e. some way to by-pass access control) which is inserted into a computer system, either accidentally or intentionnally, is a design fault; this fault may remain dormant until some malicious human makes use of it to enter the system; the intruder login is an intentional interaction fault; when the intruder is logged in (while he or she should not be), he or she may deliberately create an error, e.g. in modifying some file (integrity attack); when this file is used by an authorized user, the service will be affected, and a failure will occur.

Two additional comments, relative to the words, or labels, "fault", "error", and "failure":

a) their exclusive use in this paper does not preclude the use in special situations of words which designate, briefly and unambiguously, a specific class of impairment; this is especially applicable to faults (e.g. bug[12], defect, deficiency) and to failures (e.g. breakdown, malfunction, denial-of-service);
b) the assignment made of the particular terms fault, error, failure simply takes into account current usage: i) fault prevention, tolerance, and diagnosis, ii) error detection and correction, iii) failure rate.

Finally, it has to be stressed that the definitions given in this section are *syntactic*; accordingly, the criteria for the various classifications performed have been emphasized, and are in our view more important than the classes themselves.

5- The Means for Dependability

5.1- Dependencies between the means for dependability

All the "how to's" which appear in the basic definitions given in section 1 are in fact goals which cannot be fully reached, as all the corresponding activities are human activities, and thus imperfect. These imperfections bring in *dependencies* which explain why it is only the *combined* utilization of the

12 Including specialization of the term "bug", as in [Gra 86], which distinguishes "Heisenbugs" (for intermittent software faults, from the Heisenberg uncertainty principle) from "Bohrbugs" (for permanent software faults, "like the Bohr atom, solid, easily detected by standard techniques, and hence boring").

above methods — preferably at each step of the design and implementation process — which can best lead to a dependable computing system. These dependencies can be sketched as follows: in spite of fault prevention by means of design methodologies and construction rules (imperfect in order to be workable), faults are created. Hence the need for fault removal. Fault removal is itself imperfect, as are the off-the-shelf components — hardware or software — of the system, hence the importance of fault forecasting. Our increasing dependence on computing systems brings in the requirement for fault tolerance, which is in turn based on construction rules; hence fault removal, fault forecasting, etc. It must be noted that the process is even more recursive than it appears from the above: current computer systems are so complex that their design and implementation need computerized tools in order to be cost-effective (in a broad sense, including the capability of succeeding within an acceptable time scale). These tools themselves have to be dependable, and so on.

The preceding reasoning illustrates the close interactions between fault removal and fault forecasting, and motivates their gathering into the single term *validation*. This is despite the fact that validation is often limited to fault removal, and associated with one of the main activities involved in fault removal, verification: e.g. in "V and V" [Boe 79]; in such a case the distinction is related to the difference between "building the system right" (related to verification) and "building the right system" (related to validation)[13]. What is proposed here is simply an extension of this concept: the answer to the question "am I building the right system?" (fault removal) being complemented by "for how long will it be right?" (fault forecasting)[14]. In addition, fault removal is usually closely associated with fault prevention, forming together **fault avoidance**, i.e. how to *aim at* a fault-free system. Besides highlighting the need for validating the procedures and mechanisms of fault tolerance, considering fault removal and fault forecasting as two constituents of the same activity — validation — is of great interest in that it enables a better understanding of the notion of coverage, and thus of an important problem introduced by the above recursion: *the validation of the validation,* or how to reach confidence in the methods and tools used in building confidence in the system. **Coverage** refers here to a measure of the representativity of the situations to which the system is submitted during its validation compared to the actual

---

13 It is noteworthy that these assignments are sometimes reversed, as in the domain of communication protocols (see e.g. [Rud 85]).

14 Validation stems from "validity", which encapsulates two notions:
- validity at a given moment, which relates to fault removal;
- validity for a given duration, which relates to fault forecasting.

situations it will be confronted with during its operational life[15]. Imperfect coverage strengthens the relation between fault removal and fault forecasting, as it can be considered that the need for fault forecasting stems from imperfect coverage of fault removal.

In the sequel of this section, we examine in turns fault tolerance, fault removal and fault forecasting; fault prevention is not dealt with as it clearly relates to "general" system engineering.

5.2 - Fault tolerance

Fault tolerance [Avi 67] is carried out by error processing and by fault treatment [And 81]. **Error processing** is aimed at removing errors from the computational state, if possible before failure occurrence; **fault treatment** is aimed at preventing faults from being activated — again.

Error processing can be carried out in two ways:

- **error recovery**, where an error-free state is substituted for the erroneous state; this substitution may take on two forms [And 81]:
  - **backward recovery**, where the erroneous state transformation consists of bringing the system back to a state already occupied prior to error occurrence; this involves the establishment of **recovery points**, which are points in time during the execution of a process for which the then current state may subsequently need to be restored;
  - **forward recovery**, where the erroneous state transformation consists of finding a new state, from which the system can operate (frequently in a degraded mode);
- **error compensation**, where the erroneous state contains enough redundancy to enable the delivery of an error-free service from the erroneous (internal) state.

When error recovery is employed, the erroneous state needs to be (urgently) identified as being erroneous prior to being transformed; this is the purpose of **error detection**, hence the term of *error detection-and-recovery* which is usually employed. The association into a component of its functional processing capability together with error detection mechanisms leads to the

[15] The notion of coverage as defined here is very general; it may be made more precise by indicating its field of application, e.g.:
- coverage of a software test wrt its text, control graph, etc.
- coverage of an integrated circuit test wrt a fault model,
- coverage of fault tolerance wrt a class of faults,
- coverage of a design assumption wrt to reality.

notion of **self-checking component**, either in hardware [Car 68, Wak 78, Nic 89] or in software [Yau 75, Lap 90a]; one of the important benefits of the self-checking component approach is the ability to give a clear definition of *error confinement areas* [Sie 82]. When error compensation is performed in a system made up of self-checking components partitioned into classes executing the same tasks, then state transformation is nothing else than switching within a class from a failed component to a non-failed one, hence the corresponding approach to fault tolerance: *error detection-and-compensation*[16]. On the other hand, compensation may be applied systematically, even in the absence of errors, then providing **fault masking** (e.g. in majority vote). However, this can at the same time correspond to a redundancy decrease which is not known. So, practical implementations of masking generally involve error detection, which may then be performed *after* the state transformation.

Backward and forward error recovery are not exclusive: backward recovery may be attempted first; if the error persists, forward recovery may then be attempted. In forward recovery, it is necessary to *assess the damage* caused by the detected error, or by errors propagated before detection; damage assessment can — in principle — be ignored in the case of backward recovery, provided that the mechanisms enabling the transformation of the erroneous state into an error-free state have not been affected [And 81].

The operational time overhead necessary for error processing is radically different according to the adopted error processing form:

- in error recovery, the time overhead is longer upon error occurrence than before; especially, in backward recovery it is related to the provision of recovery points, thus in fact to preparing for error processing;
- in error compensation, the time overhead necessitated by compensation is the same, or almost the same, whether errors are present or not[17].

In addition, the duration of error compensation is much shorter than the duration of error recovery, due to the larger amount of (structural) redundancy. This remark

a) is of high practical importance in that it often conditions the choice of the adopted fault tolerance strategy with respect to the user time granularity;
b) has introduced a relation between operational time overhead and structural redundancy; more generally, a redundant system always

---

[16] Error detection-and-compensation may be seen as a limiting case of error detection-and-recovery, where recovery is performed using the present (erroneous) state of the system instead of substituting an error-free state to the erroneous state.

[17] In either case, the time for updating the system status records adds to the time overhead.

provides redundant behavior, incurring at least some operational time overhead; the time overhead may be small enough not to be perceived by the user, which means only that the *service* is not redundant; an extreme opposite form is "time redundancy" (redundant *behavior* obtained by repetition) which needs to be at least initialized by a structural redundancy, limited but existing; roughly speaking, the more the structural redundancy, the less the time overhead incurred.

The first step in *fault treatment* is **fault diagnosis**, which consists of determining the cause(s) of error(s), in terms of both location and nature. Then come the actions aimed at fulfilling the main purpose of fault treatment: preventing the fault(s) from being activated again, thus aimed at making it(them) passive, i.e. **fault passivation**. This is carried out by removing the component(s) identified as being faulty from further executions. If the system is no longer capable of delivering the same service as before, then a *reconfiguration* may take place.

If it is estimated that error processing could directly remove the fault, or if its likelihood of recurring is low enough, then fault passivation need not be undertaken. As long as fault passivation is not undertaken, the fault is regarded as a **soft fault**; undertaking it implies that the fault is considered as **hard**, or **solid.** At first sight, the notions of soft and hard faults may seem to be respectively synonymous to the previously introduced notions of temporary and permanent faults. Indeed, tolerance of temporary faults does not necessitate fault treatment, since error recovery should in this case directly remove the effects of the fault, which has itself vanished, provided that a permanent fault has not been created in the propagation process. In fact, the notions of soft and hard faults are useful due to the following reasons:

- distinguishing a permanent fault from a temporary fault is a difficult and complex task, since a) a temporary fault vanishes after a certain amount of time, usually before fault diagnosis is undertaken, and b) faults from different classes may lead to very similar errors; so, the notion of soft or hard fault in fact incorporates the subjectivity associated with these difficulties, including the fact that a fault may be declared as a soft fault when the fault diagnosis is unsuccessful;
- the ability of those notions to incorporate subtleties of the modes of action of some transient faults; for instance, can it be said that the dormant internal fault resulting from the action of alpha particles (due to the residual ionization of circuit packages), or of heavy ions in space, on memory elements (in the broad sense of the term, including flip-

flops) is a *temporary* fault? Such a dormant fault is however a *soft* fault.

The preceding definitions apply to physical faults as well as to design faults: the class(es) of faults which can actually be tolerated depend(s) on the fault hypothesis which is being considered in the design process, and thus relies on the *independency* of redundancies with respect to the process of fault creation and activation. An example is provided by considering tolerance of physical faults and tolerance of design faults. A (widely-used) method to attain fault tolerance is to perform multiple computations through multiple channels. When tolerance of physical faults is foreseen, the channels may be identical, based on the assumption that hardware components fail independently; such an approach is not suitable for the tolerance to design faults where the channels have to provide *identical services* through *separate designs and implementations* [Elm 72, Ran 75, Avi 78], i.e. through **design diversity** [Avi 84].

In dealing with fault-tolerant systems, one frequently encounters situations involving multiple faults and/or failures. Consideration of their causes leads one to distinguish:

- **independent faults**, which are attributed to different causes;
- **related faults**, which are attributed to a common cause.

Examples of common causes are (single) power supplies, clocks, specifications, etc. In design diversity, related faults may also result from dependencies in the separate designs and implementations. Related faults usually cause **common-mode failures.**

Another useful notion when discussing fault-tolerant systems is **coincident errors**, i.e. errors manifested by the same input [Avi 86].

The temporal relationship between multiple failures leads one to distinguish:

- **simultaneous failures**, which occur within some predefined time window;
- **sequential failures**, which do not occur within the same predefined time window.

Although a mathematical definition of "simultaneous" would make the time window width zero, in practice, the window may be quite wide, depending on the application, hence the notion of "near-coincident" failures which is sometimes employed [Tri 84]. Typically, in a fault-tolerant system which has been designed to tolerate a single fault at a time, it is necessary to recover from the effects of one fault before the system can tolerate the next fault. In this context, the time window that separates simultaneous from sequential faults is the interval of time necessary for error processing and possibly fault treatment, during which the system is vulnerable.

An important aspect in the coordination of the activity of multiple components is that of preventing error propagation from affecting the operation of non-failed components. This aspect becomes particularly important when a given component needs to communicate some information to other components that is private to that component. Typical examples of such *single-source information* are local sensor data, the value of a local clock, the local view of the status of other components, etc. The consequence of this need to communicate single-source information from one component to other components is that non-failed components must reach an *agreement* as to how the information they obtain should be employed in a mutually consistent way. Specific attention has been devoted to this problem in the field of distributed systems (see e.g. atomic broadcast [Cri 85a], clock synchronization [Lam 85, Kop 87] or membership protocols [Cri 88]). It is important to realize, however, that the inevitable presence of structural redundancy in any fault-tolerant system implies distribution at one level or another, and that the agreement problem therefore remains in existence. Geographically localized fault-tolerant systems may employ solutions to the agreement problem that would be deemed too costly in a "classical" distributed system of components communicating by messages (e.g. inter-stages [Lal 86], multiple stages for interactive consistency [Fri 82]).

The knowledge of some system properties may limit the necessary amount of redundancy, leading to the so-called "low-cost fault tolerance". Examples of these properties are regularities of a structural nature: error detecting and correcting codes [Pet 72], robust data structures [Tay 80], multiprocessors and computer networks [Pra 86, Ren 86], algorithm-based fault tolerance [Hua 82]. The faults which are tolerated are then dependent upon the properties accounted for, as they intervene directly in the fault hypotheses.

Of importance is the signalling of a component failure to its users. This may be accounted for within the framework of *exceptions* [Mel 77, Cri 80, And 81]. *Exception handling* facilities provided in some languages may constitute a convenient way for implementing error recovery, especially forward recovery[18].

---

[18] The use of the term "exception", due to its origin of coping with exceptional situations — not only errors— is to be carefully used in the framework of fault tolerance: it could appear as contradicting the view that fault tolerance is a natural attribute of computing systems,

Fault tolerance is (also) a recursive concept: it is essential that the mechanisms aimed at implementing fault tolerance be protected against the faults which can affect them. Examples are voter replication, self-checking checkers [Car 68], "stable" memory for recovery programs and data [Lam 81].

Fault tolerance is not restricted to accidental faults. Protection against intrusions traditionally involves cryptography [Den 82]. Some mechanisms of error detection are directed towards both intentional and accidental faults (e.g. memory access protection techniques) and schemes have been proposed for the tolerance to both intrusions and physical faults [Fra 86, Rab 89], as well as for tolerance to malicious logic [Jos 88].

5.3 - Fault removal

Fault removal is composed of three steps: verification, diagnosis, correction. **Verification** is the process of checking whether the system adheres to properties, termed the *verification conditions* [Che 81]; if it does not, the other two steps have to be undertaken: diagnosing the fault(s) which prevented the verification conditions from being fulfilled, and then performing the necessary corrections. After correction, the process has to be resumed in order to check that fault removal had no undesired consequences; the verification performed at this stage is usually termed (non-)**regression verification**. The verification conditions can take two forms:

- general conditions, which apply to a given class of systems, and are therefore — relatively — independent of the specification, e.g. absence of deadlock, conformance to design and realization rules;
- conditions specific to the considered system, directly deduced from its specification.

The verification techniques can be classed according to whether or not they involve exercizing the system. Verifying a system without actual execution is **static verification**. The verification can be conducted:

- on the system itself, in the form of a) *static analysis* (e.g. inspections or walk-through [Mye 79], data flow analysis [Ost 76], complexity analysis [McC 76], compiler checks, etc.) or b) *proof-of-correctness* (inductive assertions [Hoa 69, Cra 87]);
- on a model of the system behavior (e.g. Petri nets, finite state automata), leading to *behavior analysis* [Dia 82].

taken into consideration from the very initial design phases, and not an "exceptional" attribute.

Verifying a system through exercizing it constitutes **dynamic verification**; the inputs supplied to the system can be either symbolic in the case of **symbolic execution**, or valued in the case of verification testing, usually simply termed **testing.**

Testing exhaustively a system with respect to all its possible inputs is generally impractical. The methods for the determination of the test patterns can be classed according to two viewpoints: criteria for selecting the test inputs, and generation of the test inputs.

The *criteria* for selecting the test inputs may in turn be decomposed into three viewpoints:

- the purpose of the testing: checking whether the system satisfies its (functional) specification is **conformance testing**, whereas testing aimed at revealing faults is **fault-finding testing**;
- the system model: depending on whether the system model relates to the function or the structure of the system, leads respectively to **functional testing** and **structural testing**;
- the existence of a fault model: if there is such a fault model, **fault-based testing** [Mor 90] is conducted, aimed at revealing specific classes of faults (e.g. stuck-at-faults in hardware production [Rot 67], physical faults affecting the instruction set of a microprocessor [Tha 78], design faults in software [Goo 75, DeM 78, How 87]); if there is no fault model, the criteria may relate e.g. to path sensitization [Rap 85, Nta 88], input boundary values [Mye 79] in software, etc.[19]

The *generation* of the test inputs may be deterministic or probabilistic:

- in **deterministic testing**, test patterns are predetermined by a selective choice according to the adopted criteria,
- in **random**, or **statistical, testing**, test patterns are selected according to a defined probability distribution on the input domain; the distribution and the number of input data are determined according to the adopted criteria [Dav 81, Dur 84].

Combining the various viewpoints leads to a number of testing approaches, some of them being labelled in a condensed manner, e.g.:

- structural testing when applied to hardware generally means fault-finding, structural, fault-based, whereas it generally means fault-finding, structural, non-fault-based testing when applied to software;

---

[19] The possibility of defining a fault model is tightly related to the stage in the development process which is considered: the more advanced, the higher the possibility of defining a fault model.

- *mutation testing* of software [DeM 78] is fault-finding, structural, fault-based, deterministic testing.

Observing the test outputs and deciding whether they satisfy or not the verification conditions is known as the *oracle* problem [Adr 82]. The verification conditions may apply to the whole set of outputs or to a compact function of the latter (e.g. a system signature when testing for physical faults in hardware [Dav 86], or a "partial oracle" when testing for design faults of software [Wey 82]). When testing for physical faults, the results — compact or not — anticipated from the system under test for a given input sequence are determined by simulation [Lev 86] or from a reference system ("golden unit"). For design faults, the reference is generally the specification; it may also be a prototype, or another implementation of the same specification in the case of design diversity ("back-to-back testing", see e.g. [Bis 88]).

Some verification methods may be used in conjunction, e.g. symbolic execution may be used a) for facilitating the determination of the testing patterns [Adr 82], or b) as a proof-of-correctness method [Car 78].

As verification has to be performed all along a system's development, the above techniques apply naturally to the various forms taken by a system during its development: prototype, component, etc.

Verifying that the system cannot do *more* than what is specified is especially important with respect to intentional faults [Gas 88].

Designing a system in order to facilitate its verification is the **design for verifiability**. This is especially developed for hardware with respect to physical faults, where the corresponding techniques are then termed *design for testability* [Wil 83, McC 86].

Fault removal during the operational phase of a system's life is **corrective maintenance**, aimed at preserving or improving the system's ability to deliver a service complying with the specification[20]. Corrective maintenance can take two forms:

- **curative** maintenance, aimed at removing faults which have produced one or more errors and have been reported;

---

20 The other forms of maintenance usually distinguished are [Ram 84]:
- *adaptive maintenance*, which adjusts the system to environmental changes (e.g. change of operating systems or system data-bases);
- *perfective maintenance*, which improves the system's function by responding to customer — and designer — defined changes, which may involve removal of specification faults.

- **preventive** maintenance, aimed at removing faults before they produce errors; these faults can be:
  - physical faults having appeared since the last preventive maintenance actions,
  - design faults having led to errors in other similar systems [Ada 84].

These definitions[21] apply to non-fault-tolerant systems as well as to fault-tolerant systems, which can be maintainable on-line (without interrupting service delivery) or off-line. It is finally noteworthy that the frontier between corrective maintenance and fault treatment is relatively arbitrary; especially, curative maintenance may be considered as an — ultimate — means of achieving fault tolerance.

5.4- Fault forecasting

Fault forecasting is conducted by performing an *evaluation* of the system behavior with respect to fault occurrence or activation. Evaluation has two aspects:

- *non-probabilistic*, e.g. determining the minimal cutset or pathset of a fault tree, conducting a failure mode and effect analysis;
- *probabilistic*, aimed at determining the conformance of the system to dependability objectives expressed in terms of probabilities associated to some of the attributes of dependability, which may then be defined as *measures* of dependability.

The life of a system is perceived by its user(s) as an alternation between two states of the delivered service with respect to the specification:

- **correct service**, where the delivered service *complies with* the specification[22];
- **incorrect service**, where the delivered service *does not comply with* the specification.

A failure is thus a transition from correct to incorrect service, and the transition from incorrect to correct service is a **restoration**. Quantifying the

21 It is noteworthy that current discussions about the irrelevance of the use of the term "maintenance" when applied to software simply forget the etymology of the word: in the Middle Ages, maintenance designated the actions performed in order to keep an army battleworthy, thus including the corrective, adaptive and perfective forms of maintenance. The association of maintenance with repairing hardware is actually a (recent) deviation; associating "to maintain" with the notion of *service* would enable this etymological meaning to be revived, while at the same time removing the very source of discussion.

22 We deliberately restrict the use of "correct" to the service delivered by a system, and do not use it for the system itself: in our view, non-faulty systems hardly exist, there are only systems which may have not yet failed.

alternation of correct-incorrect service delivery enables reliability and availability to be defined as measures of dependability:

- **reliability**: a measure of the *continuous* delivery of correct service — or, equivalently, of the *time to* failure;
- **availability**: a measure of the delivery of correct service *with respect to the alternation* of correct and incorrect service.

A third measure, **maintainability**, is usually considered, which may be defined as a measure of the time to restoration from the last experienced failure, or equivalently, of the continuous delivery of incorrect service.

As a measure, safety can be seen as an extension of reliability. Let us group the state of correct service together with the state of incorrect service subsequent to benign failures into a safe state (in the sense of being free from catastrophic damage, not from danger); **safety** is then a measure of continuous safeness, or equivalently, of the time to catastrophic failure. Safety can thus be considered as reliability with respect to the catastrophic failures. A direct extension of availability, i.e. a measure of safeness with respect to the alternation of safeness and incorrect service after catastrophic failure, would not provide a significant measure. When a catastrophic failure has occurred, the consequences are generally so important that service restoration is not of prime importance for — at least — the two following reasons:

- it comes second to repairing (in the broad sense of the term, including legal aspects) the consequences of the catastrophe;
- the lengthy period prior to being allowed to operate the system again (commissions of enquiry, etc.) would lead to insignificant numerical values.

A "hybrid" reliability-availability-type measure can however be defined: a measure of correct service delivery with respect to the alternation of correct service and incorrect service after benign failure. This measure is of interest in that it provides indeed a quantification of the system availability *before* occurrence of a catastrophic failure, and as such enables quantification of the so-called "reliability- (or availability-) safety tradeoff".

In the case of multi-performing systems, several services can be distinguished, as well as several modes of service delivery, ranging from full capacity to complete disruption, which can be seen as distinguishing less and less correct service deliveries. Performance-related measures of dependability for such systems are usually termed **performability** [Mey 78, Smi 88].

The two main approaches to probabilistic fault-forecasting, aimed at deriving quantified estimates of the dependability measures, are modeling and (evaluation) testing. These approaches are directly complementary, since modeling needs data on the basic processes modeled (failure process, maintenance process, system activation process, etc.), which may be obtained by testing.

When performing an evaluation through *modeling*, the approaches differ significantly according to whether the system is considered as being in stable reliability or in reliability growth, which may be defined as follows [Lap 90]:

- **stable reliability**: the system's ability to deliver correct service is *preserved* (stochastic identity of the successive times to failure);
- **reliability growth**: the system's ability to deliver correct service is *improved* (stochastic increase of the successive times to failure)[23].

Practical interpretations of stable reliability and of reliability growth are as follows:

- stable reliability: at a given restoration, the system is identical to what it was at the previous restoration; this corresponds to the following situations:
  - in the case of a hardware failure, the failed part is changed for another one, identical and non failed,
  - in the case of a software failure, the system is restarted with an input pattern different from the one having led to failure;
- reliability growth: the fault whose activation has led to failure is diagnosed as a design fault (in software or in hardware) and is removed.

Evaluation of the dependability of systems in stable reliability is usually composed of two main phases:

- *construction* of the model of the system from the elementary stochastic processes which model the behavior of the components of the system and their interactions;
- *processing* the model in order to obtain the expressions and the values of the dependability measures of the system.

Evaluation can be conducted with respect to a) physical faults [Tri 84], b) design faults [Lit 79, Arl 88], or c) a combination of both [Lap 84, Pig 88]. The dependability of a system is highly dependent on its environment, either in

23 *Reliability decrease* (the system's ability to deliver correct service is degraded, and there is thus a stochastic decrease of the successive times to failure) is theoretically, and practically, possible, e.g. upon introduction of new faults during corrective actions, whose probability of activation is greater than for the removed fault(s). In such a case, it has to be hoped that the decrease is limited in time, and that reliability is globally growing over a long observation period of time.

the broad sense of the term [Hec 87], or more specifically its load [Cas 81, Iye 82].

Many models of reliability growth have been proposed, devoted to hardware [Dua 64], software, or both [Lap 90b]. Most of them are devoted to software, and they are aimed at evaluating either the reliability [Yam 85, Mil 86], or the number of (remaining) faults [Goe 79, Toh 89]; as these models are aimed at predicting the future reliability from the failure data accumulated in the past, particular attention has been devoted to the prediction problem [Lit 88].

Although it is not — primarily — aimed at verifying a system, *evaluation testing* [Mil 87, Cho 87] can be characterized using the viewpoints defined in section 5.3: conformance, functional, non fault-based, statistical. A major concern here is that the input profile be representative of the operational profile, hence its name: **operational testing.**

When evaluating fault-tolerant systems, the coverage of error processing and fault treatment mechanisms has a very significant influence [Bou 69, Arn 73]; its evaluation can be performed either through modeling [Dug 89] or through testing, then called *fault-injection* [Arl 89, Gun 89].

6- The Attributes of Dependability

The attributes od dependability have been defined in section 1 according to different properties, which may be more or less emphasized depending on the application intended for the computer system under consideration:

- availability is always required, although to a varying degree depending on the application;
- reliability, safety, security may or may not be required according to the application.

An additional property, which may be viewed as a prerequisite for the other properties to be fulfilled, is **integrity**, i.e. the condition of being unimpaired, in the broad sense of the term: a) for either data or programs, and b) with respect to either accidental or intentional faults.

The variations in the emphasis to be put on the attributes of dependability have a direct influence on the appropriate balance of the techniques addressed in the previous section to be employed in order that the resulting system be dependable. This is an all the more difficult problem as some of the attributes are antagonistic (e.g. availability and safety, availability and security), necessitating that trade-offs be performed. Considering the three main design dimensions of a computer system, i.e. cost, performance and dependability, the problem is further exacerbated by the fact that the dependability dimension is less understood than the cost-performance design space [Sie 82].

Maintainability was defined in section 5.4 as a measure of dependability. **Maintainability** can also be considered as a dependability attribute, relating then to the easiness with which maintenance actions can be performed.

The definition of security given in section 1, the prevention of the unauthorized access and/or handling of information, originates from [ITS 90], which defines security as the combination of confidentiality, the prevention of the unauthorized disclosure of information, integrity, the prevention of the unauthorized amendment or deletion of information, and availability, the prevention of the unauthorized withholding of information. It is worth noting that:

a) an unauthorized access or handling of information may result from either accidental or intentional faults, and that — as noted in section 5.2 — some mechanisms for protecting against unauthorized access are common to both types of faults;
b) with respect to intentional faults, the notion of authorization has to be understood broadly: an authorized individual who abuses his authority in fact violates the authorization which he was granted by performing illegitimate actions, and thus becomes an intruder.

An important feature of fault-tolerant systems is their ability, due to the presence of error detection procedures, to provide the users with information whether or not the (functional) service being delivered is correct. Such a property is termed **trustability** in [Fur 90].

The *assessment* of whether a system is truly dependable — justified reliance on the delivered service — or not thus goes beyond the validation techniques as they have been addressed in the previous sections for, at least, the three following reasons and limitations:

- checking with certainty the coverage of the design or validation assumptions with respect to reality (e.g. relevance to actual faults of the criteria used for determining test inputs, fault hypotheses in the design of fault tolerance mechanisms) would imply a knowledge and a mastering of the technology used, of the intended utilization of the system, etc. which exceeds by far what is generally achievable;

- performing an evaluation of a system according to some attributes of dependability with respect to some classes of faults is currently considered as non feasible or yielding non-significant results: probability-theoretic bases do not exist or are not — yet — widely accepted; examples are safety with respect to accidental design faults, security with respect to intentional faults;
- the specifications "against" which validation is performed are not generally non-faulty — as any system.

Among the numerous consequences of this state of affair, let us mention:
- the emphasis put on the development and production process when assessing a system: methods and techniques utilized and how they are employed; in some cases, a *grade* is assigned and delivered to the system according to a) the nature of the methods and techniques employed, and to b) an assessment of their utilization[24];
- the presence, in the specifications of some fault-tolerant systems, in addition to probabilistic requirements in terms of dependability measures, of the number of faults which are to be tolerated[25]; such a specification would not be necessary if the limitations mentioned above could be overcome.

CONCLUSION

Increasingly, individuals and organizations are developing or procuring sophisticated computing systems on whose services they need to place great reliance — whether to service a set of cash dispensers, calculate a satellite orbit, control an airplane or a nuclear plant, or to maintain the confidentiality of a sensitive data base. In differing circumstances, the focus will be on differing properties of such services — e.g. the average real-time response achieved, the likelihood of producing the required results, the ability to avoid causing failures which could be catastrophic to the system's environment, or the degree to which deliberate intrusions can be prevented. The notion of dependability provides a very convenient means of subsuming these various concerns within

---

24 For instance:
- systems are ranked from the security viewpoint [DoD 85] from A1 ("verified design") to D ("minimal protection");
- software for civil transportation airplanes are classed [RTC 85] as Level 1, 2 or 3 according to the criticality of the function to be accomplished by the software: critical, essential, non-essential.

25 Such specifications are classical in aerospace applications, under the form of a concatenation of "fail-operational" or "fail-safe" requirements.

a single conceptual framework. Dependability thus includes as special cases such properties as reliability, availability, safety, security. It also provides a means of addressing the problem that what a user usually needs from a system is an *appropriate balance* of such properties.

GLOSSARY

Warning: this glossary is provided as an aid for reading the document. Do not consider it independently of the document.

Accidental fault .... Fault appearing or created fortuitously.

Active fault .... Fault producing an error.

Arbitrary failure .... see Failure

Atomic system .... System whose internal structure cannot be discerned, or is not of interest and can be ignored.

Attributes of dependability .... Attributes enabling the system quality resulting from the impairments and the means opposing to them to be assessed.

Reliability, availability, maintainability, safety, security, trustability.

Availability .... Dependabiliy with respect to the readiness for usage.

Measure of correct service delivery with respect to the alternation of correct and incorrect service.

Avoidance (fault ~) .... Methods and techniques aimed at producing a fault-free system.

Fault prevention and fault removal.

Backward recovery .... Form of error recovery where the erroneous state transformation consists of bringing the system back to previously occupied state.

Behavior (system ~) .... What a system does.

Benign failure .... Failure whose penalties are of the same order of magnitude as the benefit provided by correct service delivery.

Catastrophic failure.................... Failure whose consequences are incommensurably greater than the benefit provided by correct service delivery.

Coincident errors........................ Errors produced on the same input.

Common-mode failures.............. Failures resulting from related faults.

Compensation (error ~)............. Form of error processing when erroneous state contains enough redundancy to enable correct service delivery.

Component (system ~)................ Another system.

Conformance testing.................. Testing whose purpose is checking whether the system satisfies its specification.

Consistent failure........................ Failure perceived similarly by all system users.

Correct service .... Service delivered in compliance with the system specification.

Corrective maintenance .... Preservation or improvement during its operational life of a system's ability to deliver a service complying with the specification .

Fault removal during the operational life of a system.

Coverage...................................... Measure of the representativity of the situations to which a system is submitted during its validation compared to the actual situations it will be confronted with during its operational life.

Crash failure................................ Persistent omission failure.

Criticality (system ~)................. Highest severity of failure modes.

Curative maintenance .... Corrective maintenance aimed at removing faults which have produced errors and which have been reported.

Dependability.............................. Trustworthiness of a computer system such that reliance can justifiably be placed on the service it delivers.

Design diversity .... An approach to the production of systems, involving the provision of identical services from separate designs and implementations.

Design fault .... Human-made internal fault.

Design for verifiability............. Methods and techniques when designing a system which facilitate its verification.

Detection (error ~) .... The action of identifying that a system state is erroneous.

Detected error .... Error recognized as such by a detection algorithm or mechanism.

Deterministic testing .... Form of testing where the test patterns are predetermined by a selective choice.

Diagnosis (fault ~) .... The action of determining the cause of an error in location and nature.

Dormant fault .... Internal fault not activated by the computation process.

Dynamic verification .... Verification involving exercizing the system.

Environment (system ~) .... The other systems having interacted or interfered, interacting or interfering, or likely to interact or interfere with the considered system.

Error .... Part of system state which is liable to lead to failure.

Manifestation of a fault in a system.

External fault .... Fault resulting from environmental interference or interaction.

Fail-safe system .... System whose failures can only be, or are to an acceptable extent, benign failures.

Fail-silent system .... System whose failures can only be, or are to an acceptable extent, crash failures.

Fail-stop system .... System whose failures can only be, or are to an acceptable extent, stopping failures.

Failure .... Deviation of the delivered service from compliance with the specification.

Transition from correct service delivery to incorrect service delivery.

Fault .... Adjudged or hypothesized cause of an error.

Error cause which is intended to be avoided or tolerated.

Consequence for a system of the failure of another system which has interacted or is interacting with the considered system.

Fault-based testing .... Testing aimed at revealing specific classes of faults.

Fault-finding testing .... Testing whose purpose is revealing faults.

Forecasting (fault ~) .... Methods and techniques aimed at estimating the present number, the future incidence, and the consequences of faults.

Forward recovery .... Form of error recovery where the erroneous state transformation consists of finding a new state.

Function (system ~) .... What a system is intended for.

Functional testing .... Form of testing where the testing inputs are selected according to criteria relating to the system's function.

Hard fault or **solid fault** .... Fault necessitating passivation.

Human-made fault .... Fault resulting from human imperfection.

Impairments to dependability .... Undesired, but not unexpected, circumstances causing or resulting from undependability.

Faults, errors, and failures.

Inconsistent failure .... Failure such that system users may have different perceptions of it.

Incorrect service .... Service delivered not in compliance with the system specification.

Independent faults .... Faults attributed to different causes.

Integrity .... Condition of being unimpaired.

Intentional fault .... Fault created deliberately.

Intermittent fault .... Temporary internal fault.

Faults whose conditions of activation cannot be reproduced or which occur rarely enough.

Internal fault .... Fault inside a system.

Intrusion .... Intentional operational external fault.

Latent error .... Error not recognized as such.

Maintainability .... Easiness with which maintenance actions can be performed.

Measure of continuous incorrect service delivery.

Measure of the time to restoration from the last experienced failure.

Malicious logic .... Intentional design fault.

Masking (fault ~) .... The result of applying error compensation systematically, even in the absence of error.

Means for dependability............Methods and techniques enabling a) to provide a system with the ability to deliver a service on which reliance can be placed, and b) to reach confidence in this ability.

Fault prevention, fault tolerance, fault removal, fault forecasting.

Omission failure..........................Failure such that no service is delivered.

Operational fault.........................Faults which appear during the system's exploitation.

Operational testing......................Testing performed for evaluating a system's dependability, with an input profile representative of its operational conditions.

Passivation (fault ~)...................The actions taken in order that a fault cannot be activated.

Performability............................Performance-related measure of dependability.

Permanent fault..........................Fault whose presence is not related to pointwise conditions of the system, either internal or external.

Physical fault..............................Fault resulting from adverse physical phenomena.

Preventive maintenance ....Corrective maintenance aimed at removing faults before they are activated.

Prevention (fault ~)...................Methods and techniques aimed at preventing fault occurrence or introduction.

Processing (error ~)..................The actions taken in order to eliminate errors from a system.

Procurement of dependability...Methods and techniques intended to provide a system with the ability to deliver a service complying with the specification.

Fault prevention and fault tolerance.

Random testing...........................See Statistical testing

Real-time function.....................Function required to be fulfilled within finite time intervals dictated by the environment.

Real-time service.......................Service required to be delivered within finite time intervals dictated by the environment.

Real-time system........................System fulfilling at least one real-time function or delivering at least one real-time service.

Recovery (error ~)........................Form of error processing where an error-free state is substituted for an erroneous state.

Recovery point............................Point in time during the execution of a process for which the then current state may subsequently need to be restored.

Regression verification..............Verification performed after a correction, in order to check that the correction has no undesired consequences.

Related faults...............................Faults attributed to a common cause.

Reliability ....Dependability with respect to the continuity of service.

Measure of continuous correct service delivery.

Measure of the time to failure.

Reliability growth.......................The system's ability to deliver correct service is improved (stochastic increase of the successive times to failure).

Removal (fault ~)........................Methods and techniques aimed at reducing the presence (number, seriousness) of faults.

Restoration (service ~)...............Transition from incorrect to correct service delivery.

Safety ....Dependability with respect to the non occurrence of catastrophic failures.

Measure of continuous delivery of either correct service or incorrect service after benign failure.

Measure of the time to catastrophic failure.

Security ....Dependability with respect to the prevention of unauthorized access and/or handling of information.

Self-checking component ....Component comprising error detection mechanisms associated with its functional part.

Sequential failures......................Failures which do not occur within the same predefined time window.

Service ....System behavior as perceived by the system user.

Severity (failure ~)....................Grade of the failure consequences upon the system environment.

Simultaneous failures.................Failures which occur within some predefined time window.

Soft fault.....................................Fault for which fault passivation is not undertaken.

Solid fault...................................see hard fault

Specification (system ~)...........Agreed description of the system's requirements.

State (system ~)..........................A condition of being with respect to a set of circumstances.

Stable reliability.........................The system's ability to deliver correct service is preserved (stochastic identity of the successive times to failure).

Static verification.......................Verification conducted without exercizing the system.

Statistical testing.........................Form of testing where the test patterns are selected according to a defined probability distribution on the input domain.

Stopping failure..........................System activity, if any, is not any more perceptible to the users, and a constant value service is delivered.

Structure (system ~)..................What makes a system do what it does.

Structural testing.......................Form of testing where the testing inputs are selected according to criteria relating to the system's structure.

Symbolic execution....................Dynamic verification performed with symbolic inputs.

System...Entity having interacted, interacting, or able to interact with other entities.

Set of components bound together in order to interact.

Temporary fault..........................Fault which is present for a limited amount of time.

Testing...Dynamic verification performed with valued inputs.

Timing failure............................Failure such that the timing of service delivery does not comply with the specification.

Tolerance (fault ~).....................Methods and techniques aimed at providing a service complying with the specification in spite of faults.

Transient fault............................Temporary physical external fault.

Treatment (fault ~) .... The actions taken in order to prevent a fault from being re-activated.

Trustability .... System's ability to provide users with information about service correctness.

User (system ~) .... Another system (physical, human) interacting with the considered system.

Validation (dependability ~) ...... Methods and techniques intended to enable confidence to be reached in a system's ability to deliver a service complying with the specification.

Fault removal and fault forecasting.

Value failure .... Failure such that the value of the delivered service does not comply with the specification.

Verification .... The process of determining whether a system adheres to properties (the verification conditions) which can be a) general, independent of the specification, or b) specific, deduced from the specification.

SURETE DE FONCTIONNEMENT:

CONCEPTS DE BASE

ET TERMINOLOGIE

INTRODUCTION

Ce document a pour objectif de donner des définitions informelles mais cependant précises des différents attributs de la sûreté de fonctionnement des systèmes informatiques. Il s'agit d'une contribution aux travaux entrepris au sein de la communauté scientifique et technique "Sûreté de Fonctionnement Informatique et Tolérance aux Fautes" [Avi 67, Jes 77, Mel 77, Avi 78, Ran 78, Car 79, And 81, FTC 82, Sie 82, Cri 85a, Lap 85, Avi 86, Lap 89], afin de proposer des définitions de certains concepts de base qui soient claires et d'une large acception.

La sûreté de fonctionnement est d'abord introduite en tant que concept générique qui englobe les notions habituelles de fiabilité, disponibilité, sécurité. Les définitions de base de la première partie sont ensuite commentées, et complétées par d'autres définitions dans les parties suivantes. Un glossaire récapitulant les définitions est donné en annexe. La présentation a été structurée de façon à éviter des références postérieures. Des caractères gras sont utilisés lorsqu'un terme est défini, l'utilisation d'italiques étant destinée à attirer l'attention du lecteur. Les lignes directrices qui ont guidé la rédaction peuvent être résumées comme suit :

- recherche d'un nombre réduit de concepts permettant d'exprimer les attributs de la sûreté de fonctionnement ;
- utilisation de termes identiques ou aussi proches que possible de ceux d'usage courant ; un terme non défini doit être compris dans sa signification habituelle (telle que fournie par un dictionnaire) ;
- accent mis sur l'intégration des concepts [Gol 82, Ran 86] (par opposition à la spécialisation) grâce à des définitions indépendantes des classes de fautes.

Ce document peut être vu comme un consensus minimum au sein de la communauté afin de faciliter des interactions fructueuses ; de plus, ce document est susceptible d'être utilisé par d'autres organisations (y compris les comités de normalisation), ainsi qu'à des buts éducatifs. L'effort terminologique n'est pas une fin en soi : les mots ne sont intéressants qu'en tant qu'étiquettes non-ambiguës attachées à des concepts, et qu'en tant que vecteurs permettant de dialoguer et de partager des idées. Ce document est dénué de toute prétention d'état de l'art ou de "Tablettes de la Loi" : les concepts qui sont présentés se doivent d'évoluer avec la technologie, et avec nos progrés dans la compréhension et la maîtrise de la spécification, de la conception et de l'évaluation de la sûreté de fonctionnement des systèmes informatiques.

Ce qui est exposé dans ce document n'existerait pas sans les nombreuses discussions avec de nombreux collègues, tout particulièrement les membres du WG 10.4 de l'IFIP.

1- DÉFINITIONS DE BASE

La **sûreté de fonctionnement** d'un système informatique est la propriété qui permet à ses utilisateurs de placer une confiance justifiée dans le service qu'il leur délivre [Car 82]. Le **service** délivré par un système est son comportement *tel que perçu* par son, ou ses, utilisateurs ; un **utilisateur** est un autre système (humain ou physique) qui *interagit* avec le système considéré.

Selon la, ou les applications auxquelles le système est destiné, l'accent peut être mis sur différentes facettes de la sûreté de fonctionnement, ce qui revient à dire que la sûreté de fonctionnement peut être vue selon des propriétés différentes mais complémentaires, qui permettent de définir ses *attributs* :

- par rapport au fait d'être *prêt à l'utilisation*, la sûreté de fonctionnement est perçue comme la **disponibilité**,
- par rapport à la *continuité du service*, la sûreté de fonctionnement est perçue comme la **fiabilité**,
- par rapport à la *non-occurrence de défaillances catastrophiques*, la sûreté de fonctionnement est perçue comme la **sécurité**-innocuité,
- par rapport à la *prévention d'accès ou de manipulations non-autorisées de l'information*, la sûreté de fonctionnement est perçue comme la **sécurité**-confidentialité[1].

Une **défaillance** du système survient lorsque le service délivré n'est plus conforme à la spécification, la **spécification** étant une description agréée de la fonction ou du service attendu du système. Une **erreur** est la partie de l'état du système qui est susceptible d'entraîner une défaillance : une erreur affectant

1 Nous sommes conscients de l'ambiguïté résultant de l'utilisation du même terme, sécurité, pour désigner deux notions différentes (mais non indépendantes). Nous ne faisons que nous conformer à l'usage établi, tout en notant que réserver, par exemple, sûreté pour la non-occurrence de défaillances catastrophiques, et sécurité pour la prévention d'accès ou de manipulations non-autorisées de l'information, ne résoudrait qu'imparfaitement le problème puisque ces deux termes partagent le même adjectif : sûr. C'est ce qui nous a conduits à associer un qualificatif pour lever l'ambiguïté : innocuité pour l'évitement des défaillances catastrophiques, confidentialité pour la prévention d'accès ou de manipulations non-autorisées de l'information. Ce dernier qualificatif a été choisi en accord avec [Gas 88] qui note que la confidentialité est la propriété la plus distinctive de la sécurité, les deux autres propriétés étant l'intégrité et le non-refus de service, c'est-à-dire la disponibilité. Naturellement, lorsque le contexte permet de lever aisément le doute, ces qualificatifs ne sont pas nécessaires.

le service est une indication qu'une défaillance survient ou est survenue. La cause adjugée ou supposée d'une erreur est une **faute**.

Le développement d'un système sûr de fonctionnement passe par l'utilisation *combinée* d'un ensemble de méthodes qui peuvent être classées en :

- **prévention des fautes** : comment empêcher l'occurrence ou l'introduction de fautes,
- **tolérance aux fautes** : comment fournir un service conforme à la spécification en dépit des fautes,
- **élimination des fautes** : comment réduire la présence (nombre, sévérité) des fautes,
- **prévision des fautes** : comment estimer la présence, la création et les conséquences des fautes.

Prévention des fautes et tolérance aux fautes peuvent être vues comme constituant l'**obtention** de la sûreté de fonctionnement : comment *fournir* au système l'aptitude à délivrer un service conforme à la spécification ; élimination des fautes et prévision des fautes peuvent être vues comme constituant la **validation** de la sûreté de fonctionnement : comment *avoir confiance* dans l'aptitude du système à délivrer un service conforme à la spécification.

Les notions qui ont été introduites peuvent être groupées en trois classes (figure 1) :

- les **entraves** à la sûreté de fonctionnement : fautes, erreurs, défaillances; elles sont les circonstances indésirables — mais non inattendues — causes ou résultats de la non-sûreté de fonctionnement (dont la définition se déduit simplement de celle de la sûreté de fonctionnement : la confiance ne peut plus, ou ne pourra plus, être placée dans le service délivré) ;
- les **moyens** pour la sûreté de fonctionnement : prévention des fautes, tolérance aux fautes, élimination des fautes, prévision des fautes ; il s'agit des méthodes et techniques permettant de fournir au système l'aptitude à délivrer un service conforme à la spécification, et de donner confiance dans cette aptitude ;
- les **attributs** de la sûreté de fonctionnement : disponibilité, fiabilité, sécurité-innocuité, sécurité-confidentialité ; ils permettent a) d'exprimer les propriétés qui sont attendues du système, et b) d'apprécier la qualité du service délivré, telle que résultant des entraves et des moyens de s'y opposer.

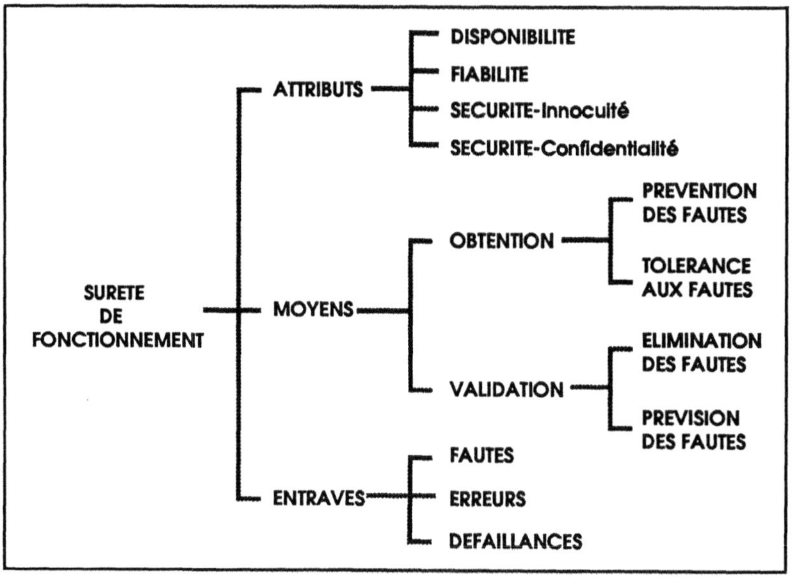

Figure 1 - L'arbre de la sûreté de fonctionnement

2- Sur l'introduction de la surete de fonctionnement en tant que concept generique

Une tendance naturelle de toute discipline scientifique ou technique qui se développe est, dans un premier temps, de restreindre son champ d'investigation afin de pouvoir progresser rapidement. Vient ensuite le temps où les interactions avec d'autres disciplines ne peuvent plus être ignorées. La tentation est alors grande de déclarer que ces autres disciplines sont des "cas particuliers" de la discipline considérée. Ceci a généralement pour conséquence nombre de débats, souvent conduits par les spécialistes de chaque discipline dans leur jargon propre. Tel fut le cas pour la fiabilité, la sécurité-innocuité et la sécurité-confidentialité. Le premier souci fut de faire fonctionner les systèmes informatiques : *fiabilité*. Les systèmes informatiques devenant fiables, leurs services sont utilisés et deviennent nécessaires pour une utilisation régulière, et la *disponibilité* devient essentielle. Leur utilisation dans des applications critiques amène le souci de sécurité-innocuité. Un système qui ne fait rien ne peut avoir de défaillances catastrophiques, mais n'est pas très utile ; d'où la prétention de certains cercles de la sécurité-innocuité de considérer la fiabilité comme un sous-ensemble de leur spécialité. L'avènement des systèmes informatiques répartis attire l'attention sur les fautes intentionnelles ; ces dernières peuvent avoir des conséquences catastrophiques, d'où la répétition du phénomène précédemment mentionné : les spécialistes de la sécurité-confidentialité considèrent généralement fiabilité et sécurité-innocuité comme des sous-ensembles de leur spécialité.

En fait, les relations entre fiabilité, sécurité-innocuité et sécurité-confidentialité sont plus complexes qu'une simple dépendance. Il n'est pour s'en convaincre que de considérer l'exemple des "bombes logiques", c'est-à-dire de fautes délibérément introduites dans un système informatique afin de provoquer, à un moment choisi par le "terroriste" — et sous son contrôle — une défaillance du système, de conséquences de préférence non perçues par l'utilisateur (jusqu'à ce qu'il devienne conscient des causes de la défaillance). Dans cet exemple, fiabilité, sécurité-innocuité et sécurité confidentialité sont de toute évidence étroitement imbriquées, et de manière variable selon le point de vue considéré[2]. Ce qui est certain, c'est que l'utilisateur ne peut, ou ne devrait pas, avoir confiance dans le service délivré par un tel système, qui n'est pas *sûr de fonctionnement.*

Ce qui précède montre clairement qu'il n'est pas dans les intentions de ce document de contribuer à la controverse de savoir si la fiabilité est un sous-ensemble de la sécurité-innocuité ou vice-versa, et de même lorsque l'on considère en plus la sécurité-confidentialité. Ce qui importe réellement dans les relations entre fiabilité, sécurité-innocuité, sécurité-confidentialité et la notion de sûreté de fonctionnement, est que les trois premières sont des *attributs* de cette dernière, d'où un enrichissement réciproque. Telle est la raison principale pour l'adjonction d'une notion supplémentaire à une liste déjà longue — fiabilité, disponibilité, sécurité, ... Une autre raison pour l'introduction de la sûreté de fonctionnement en tant que concept générique est la volonté d'affranchir fiabilité de sa signification étymologique globale pour se concentrer sur sa signification largement répandue (et historiquement récente) de continuité de service[3].

---

2 Il est également à noter que les événements relatés dans la partie "Risk to the Public in Computer Systems" du périodique ACM Software Engineering Notes sont relatifs tant à la fiabilité qu'à la sécurité-inocuité et à la sécurité-confidentialité.

3 Il est intéressant de noter que :

a) la plupart des ouvrages comportant le mot "fiabilité" dans leur titre traitent d'évaluation, de mesure, de prévision de la fiabilité des systèmes, et non de la construction de systèmes fiables ;

b) le fait de voir la sûreté de fonctionnement comme un concept plus général que fiabilité, disponibilité, ... et incorporant ces notions, n'est pas nouveau (voir par exemple [Hos 60]) ; cependant, ces essais passés ne comportaient pas la généralité de notre approche, puisque le but était de définir une mesure englobant disponibilité et fiabilité, et que la sécurité-confidentialité n'était pas considérée.

D'un point de vue étymologique, la notion de fiabilité était déjà présente au douzième siècle sous le nom de "fiableté", dont la signification était "caractère de ce qui est digne de confiance" ; l'origine latine est le verbe "fidare", qui dans le langage populaire signifiait "avoir confiance". A la lumière de ces considérations étymologiques, on ne peut que regretter que la définition de fiabilité habituellement utilisée dans l'ingénierie des systèmes[4] ait substitué la notion d'aptitude à la notion de confiance, et ce pour (au moins) les deux raisons suivantes :

a) du point de vue de l'utilisateur d'un système, ce qui l'intéresse réellement n'est pas tant *l'aptitude* à remplir des fonctionnalités, que le *service* qui lui est *effectivement* délivré ;
b) du point de vue de producteur du système qui est prêt à admettre l'existence possible de fautes dans sa conception, l'interprétation du terme "aptitude" est sujette à caution, bien qu'ayant été adoptée dans les glossaires relatifs au génie logiciel (voir par exemple [IEE 82]).

3- Sur la fonction, le comportement, la structure et la spécification d'un systeme

Jusqu'à présent, un **système** a été – implicitement – considéré comme un tout, l'accent étant mis sur son comportement tel que perçu de l'extérieur. Une définition selon cette vue "boîte noire" est : un système est une entité ayant interagi ou interféré, interagissant ou interférant, ou susceptible d'interagir ou d'interférer avec d'autres entités, c'est-à-dire d'autres systèmes. Ces autres systèmes ont constitué, constituent, ou constitueront l'**environnement** du système considéré[5]. Un utilisateur du système est une partie de l'environnement qui *interagit* avec ce dernier : un utilisateur fournit des entrées au système et/ou en reçoit des sorties ; ce qui le distingue du reste de l'environnement est le fait qu'il *utilise le service* délivré par le système.

---

[4] Par exemple, «Fiabilité : aptitude d'un dispositif à accomplir une fonction requise dans des conditions données, pour un intervalle de temps donné» [IEC 85].

[5] a) Le fait de donner des définitions récursives n'est pas pour le plaisir de la récursion. Le but est de souligner la relativité de la notion même de système selon le point de vue adopté : un système ne sera pas vu de la même façon par ses concepteurs, ses utilisateurs, ses équipes de maintenance.
b) L'emploi du passé, du présent et du futur a pour objectif de mettre l'accent sur le fait que l'environnement d'un système peut évoluer, en particulier avec les phases de son cycle de vie. Par exemple, la notion d'"environnement de programmation" s'intègre dans la définition donnée, tout autant que l'environnement physique auquel un système est confronté dans sa vie opérationnelle.

La **fonction** d'un système est ce à quoi le système *est destiné* [Kui 85]. Le **comportement** d'un système est ce qu'il *fait*. Ce qui *lui permet de faire ce qu'il fait* est sa **structure** [Zie 76]. En adoptant l'esprit de [And 81], une définition d'un système d'un point de vue structurel ('boîte blanche" ou "boîte de verre") est la suivante : un système est un ensemble de composants interconnectés en vue d'interagir ; un **composant** est un autre système, etc. La décomposition s'arrête lorsqu'un système est considéré comme étant un **système atomique** : aucune décomposition ultérieure n'est envisagée, soit par nature, soit parce que dénuée d'intérêt. Le terme "composant" doit être compris dans un sens large : couches d'un système autant que composants intra-couche ; de plus, un composant étant lui-même un système, il englobe les relations entre les composants qui le constituent. Une définition plus classique de la structure d'un système est ce que le système est. Cette définition convient tant que l'on ne s'intéresse pas aux entraves à la sûreté de fonctionnement, et donc lorsque l'on considère la structure comme figée. Nous ne voulons pas nous limiter aux composants dont la structure est figée : en fait, les entraves à la sûreté de fonctionnement peuvent être, causer, ou résulter de changements structurels. Une structure peut donc avoir des états[6]. D'où une définition de la notion d'état : un **état** est une condition d'être *par rapport à un ensemble de circonstances* ; cette définition s'applique tant au comportement d'un système qu'à sa structure[7].

De par sa définition (le comportement tel que perçu par un utilisateur), le service délivré par un système est clairement une *abstraction* de son comportement. Il est à noter que cette abstraction dépend directement de l'application pour laquelle est utilisé le système. Un exemple de cette dépendance est le rôle joué par le temps dans cette abstraction : les granularités temporelles d'un système et de ses utilisateurs sont généralement différentes, et varient d'une application à l'autre. De plus, la notion de service n'est bien sûr pas restreinte aux sorties exclusivement, mais englobe toutes les interactions qui intéressent l'utilisateur ; par exemple, balayer des capteurs fait clairement partie du service attendu d'un système de surveillance.

---

6 a) On peut donc dire qu'une structure a également un comportement, tout particulièrement par rapport aux entraves à la sûreté de fonctionnement, même si les vitesses d'évolution considérées par rapport d'une part aux requêtes d'un utilisateur, et d'autre part aux entraves à la sûreté de fonctionnement sont — il faut le souhaiter — radicalement différentes.

b) La définition donnée permet d'englober d'autres types de systèmes à structure variable, par exemple les systèmes adaptatifs, et particulièrement les systèmes basés sur la connaissance.

7 Cette définition est destinée à insister sur la relativité de la notion d'état, qui dépend directement des phénomènes et circonstances considérés ; par exemple : états par rapport aux activités de traitement des informations; états par rapport à l'occurrence des défaillances.

Nous avons jusqu'à présent utilisé le singulier pour fonction et service. Un système remplit généralement plus d'une fonction, et délivre plus d'un service. Fonction et service peuvent être donc vus comme constitués d'éléments de fonction et d'éléments de service. Dans un but de simplicité, nous utiliserons simplement le pluriel — fonctions, services — lorsqu'il est nécessaire ou utile de distinguer plusieurs éléments de fonction ou de service.

Les propriétés d'un système relatives au respect de contraintes temporelles sont d'un intérêt particulier pour la sûreté de fonctionnement. Une **fonction** ou un **service temps réel** est une fonction qui doit être remplie ou un service qui doit être délivré dans des intervalles de temps finis *régis par l'environnement* ; un **système temps réel** est un système qui remplit au moins une fonction temps réel ou qui délivre au moins un service temps réel [PDC 90].

La spécification d'un système décrit ce qui est attendu d'un système en termes a) de sa fonction ou de son service ou des deux, et b) des conditions dans — ou selon — lesquelles la remplir ou le délivrer : environnement, durée, performance, observabilité, ... La fonction et/ou le service sont habituellement spécifiés d'abord en termes de ce qui devrait être rempli ou délivré concernant la finalité première du système. Lorsque l'on considère des systèmes de sécurité (soit -innocuité, soit -confidentialité), cette spécification est généralement complétée par l'édiction de ce qui ne devrait pas arriver (par exemple, les états dangereux pouvant mener à une catastrophe, ou la divulgation d'informations sensibles). Cette dernière spécification conduit généralement à spécifier des fonctions ou services additionnels que le système devrait remplir ou délivrer afin de réduire les possibilités de ce qui ne devrait pas arriver (par exemple, vérification des droits d'un utilisateur et son identification).

De plus, ces diverses spécifications peuvent :

a) être exprimées selon divers degrés de détail : spécification des besoins, spécification de conception, spécification de réalisation, ...
b) être décomposées selon l'absence ou la présence de défaillance ; le premier cas est relatif à ce qui est habituellement appelé le mode d'opération *nominal*, et le second peut être relatif à un mode d'opération *dégradé*, si les ressources survivantes ne sont plus suffisantes pour assurer le mode nominal.

En conséquence, il n'y a pas généralement une seule spécification, mais plusieurs, et un système peut défaillir par rapport à l'une de ces multiples spécifications, et satisfaire encore les autres.

Il est essentiel qu'une spécification soit agréée par deux personnes, ou groupes de personnes, physiques ou morales : le fournisseur du système (au sens large du terme : concepteur, constructeur, vendeur, ...) et ses utilisateurs humains[8]. L'agrément est nécessaire pour que la spécification serve de base pour déterminer si le service délivré est acceptable ou non. Ce qui peut être jugé comme un service acceptable par rapport à une spécification à un niveau de détail donné peut ne pas satisfaire cette spécification à un autre niveau de détail, en raison de méprises commises en détaillant la spécification, résultant en fait en *fautes de spécification* ; des fautes de spécification peuvent à leur tour affecter n'importe laquelle des diverses spécifications. Plus généralement, une spécification ne peut prétendre être immuable une fois établie. Ce serait simplement ignorance de la vie, qui est *changement*. Les changements peuvent être motivés par la modification des besoins du système : modification de la fonction et/ou du service attendus, ou correction de fautes[9]. Une fois encore, ce qui est important est que la spécification fasse — à nouveau — l'objet d'un agrément.

Du fait de la définition donnée ci-dessus pour la structure d'un système, les notions de fonction, de service et de leur spécification s'appliquent naturellement aux composants. Ceci revêt un intérêt particulier dans le processus de conception, lorsque des composants pré-existants, tant matériels que logiciels sont incorporés dans un système : ce qui intéresse davantage le concepteur est la fonction ou le service délivré par le composant, plus que son comportement (interne) détaillé.

4- Les entraves à la sureté de fonctionnement

4.1- Fautes

Les fautes et leurs sources sont extrêmement diverses. Les trois points de vue principaux selon lesquels ont peut les classer sont leur nature, leur origine et leur persistance.

---

8 L'agrément peut être implicite, comme lorsque l'on achète un système accompagné de sa spécification et de son manuel d'utilisation, ou lorsque l'on emploie des systèmes "sur étagère".

9 Nous sommes donc confrontés à un problème circulaire : une référence est nécessaire pour déterminer si un service délivré est acceptable ou non, et cette référence peut être fautive. Une attention particulière a depuis longtemps été portée à l'amélioration des spécifications, y compris des propositions de modèles de cycle de vie ayant cet objectif [Boe 88].

La *nature* des fautes conduit à distinguer :

- les **fautes accidentelles**, qui apparaissent ou sont créées de manière fortuite,
- les **fautes intentionnelles**, qui sont créées délibérément, avec une intention qui peut être présumée nuisible.

L'*origine* des fautes regroupe elle-même trois points de vue :

- la *cause phénoménologique*, qui conduit à distinguer [Avi 78] :
  - les **fautes physiques**, qui sont dues à des phénomènes physiques adverses,
  - les **fautes humaines**, qui résultent d'imperfections humaines ;
- les *frontières du système*, qui conduisent à distinguer :
  - les **fautes internes**, qui sont les parties de l'état d'un système qui, lorsqu'activées par les traitements, produiront une ou des erreurs,
  - les **fautes externes**, qui résultent de l'interférence ou des interactions du système avec son environnement physique (perturbations électro-magnétiques, radiations, température, vibrations, ...) ou humain ;
- la *phase de création* par rapport à la vie du système, qui conduit à distinguer :
  - les **fautes de conception**, qui résultent d'imperfections commises soit au cours du développement du système (de l'expression des besoins à la recette, y compris l'établissement des procédures d'exploitation ou de maintenance), soit au cours de modifications ultérieures,
  - les **fautes opérationnelles**, qui apparaissent durant l'exploitation du système.

La *persistance temporelle* conduit à distinguer :

- les **fautes permanentes**, dont la *présence* n'est pas reliée à des conditions ponctuelles, internes (processus de traitement) ou externes (environnement),
- les **fautes temporaires**, qui sont reliées à de telles conditions, et qui sont donc présentes pour une durée limitée.

Les problèmes de sécurité-confidentialité sont dominés par les fautes intentionnelles, mais non limités à ces dernières, qui sont clairement des fautes *humaines*. Les fautes intentionnelles peuvent être internes ou externes ; des exemples typiques sont :

- en ce qui concerne les fautes internes, l'incorporation de **logique maligne** (par exemple des "chevaux de Troie"), qui est une faute de *conception* intentionnelle ;
- en ce qui concerne les fautes externes, une **intrusion** qui est une faute *opérationnelle* intentionnelle.

Afin d'être "couronnées de succès", les fautes intentionnelles peuvent profiter de fautes accidentelles, par exemple une intrusion exploitant une brêche de sécurité due à une faute de conception accidentelle. On peut tracer des similarités évidentes et néammoins intéressantes entre cet exemple et une faute accidentelle temporelle externe "exploitant" un défaut de blindage.

Le fait de considérer la cause phénoménologique des fautes peut conduire récursivement "loin en arrière", par exemple pourquoi les programmeurs font-ils des erreurs? pourquoi les circuits intégrés défaillent-ils? La notion même de faute est *arbitraire*, et constitue en fait une facilité permettant de mettre un point d'arrêt à la récursion. D'où la définition donnée d'une faute : cause *adjugée ou supposée* d'une erreur. Cette cause peut varier en fonction du point de vue adopté : mécanismes de tolérance aux fautes, équipes de maintenance, concepteurs, physiciens du solide, etc. La récursion s'arrête à *la cause que l'on désire prévenir ou tolérer*. Ceci procure une cohérence à la distinction entre faute humaine et faute physique : un système informatique étant une création humaine, toute faute l'affectant est de façon ultime due aux humains en ce sens qu'elle est la conséquence de notre inaptitude à maîtriser tous les phénomènes qui régissent le comportement d'un système. De manière absolue, la distinction entre fautes humaines et fautes physiques peut donc être considérée comme non-nécessaire ; cependant, cette distinction est importante à la lumière des méthodes actuelles pour l'obtention et la validation de la sûreté de fonctionnement. Si la récursion précédemment mentionnée n'est pas arrêtée, alors *une faute n'est autre que la conséquence de la défaillance d'un système qui a délivré (y compris les concepteurs) ou qui délivre un service au système considéré.*

Des exemples de ce qui précède:

- une faute de conception est consécutive à une défaillance d'un concepteur ;
- une faute physique interne est due à une défaillance d'un composant matériel, qui est elle-même la conséquence d'une ou plusieurs erreurs au niveau électrique ou électronique, ayant à son tour sa source dans le processus de production du matériel, ou dans les limites de nos connais-

sances en physique du solide ; les fautes physiques internes peuvent être qualifiées de fautes *récurrentes* tant que leur réparation consiste à remplacer le composant défaillant par un composant identique — non-défaillant : le nouveau composant peut défaillir à nouveau de manière similaire sauf si la cause de la défaillance a été identifiée dans le processus de conception et de production, et éliminée, conduisant ainsi à un composant modifié dans le but de réduire sa probabilité de défaillance ;

- une faute externe, qu'elle soit physique ou humaine, est en fait une faute de conception : l'inaptitude à prévoir toutes les situations auxquelles un système sera confronté durant sa vie opérationnelle, ou le refus conscient (par exemple pour des raisons de compromis économique) de considérer certaines d'entre elles ; par exemple :
  - une perturbation électromagnétique affectant l'état interne d'un système est-elle une faute physique externe ou le résultat d'une faute de conception, c'est-à-dire un blindage inadéquat?
  - l'action d'un opérateur tapant au clavier un caractère inapproprié qui conduit à défaillance est elle une faute d'interaction ou une faute de conception, c'est-à-dire l'absence de confirmation demandée par le système [Nor 83]?

Deux commentaires en ce qui concerne les fautes temporaires :

1) Les fautes temporaires externes ayant leur source dans l'environnement physique sont souvent dénommées **fautes transitoires.**
2) Les fautes temporaires internes sont souvent dénommées **fautes intermittentes.** Ces fautes résultent de la présence de combinaisons survenant rarement, telles que :
   - fautes sensibles aux entrées dans les mémoires à semi-conducteurs, changements dans les paramètres d'un composant matériel (effet de variations de température, retard dans les temporisations dû à des capacités parasites) ;
   - situations – affectant aussi bien le logiciel que le matériel – survenant quand la charge excède une certaine valeur, par exemple des temporisations ou des synchronisations marginales.

   La notion même de faute intermittente est, dans l'absolu, arbitraire : de telles fautes ne sont autres que des fautes — permanentes — dont les conditions d'activation ne peuvent être reproduites ou qui se manifestent suffisamment rarement ; cependant, comme déjà mentionné à propos de la distinction entre fautes physiques et fautes humaines, leur accorder existence est utile.

Il apparaît de ce qui précède que *toute faute peut être considérée comme une faute de conception permanente*. Ceci est effectivement vrai de façon ultime, mais n'est probablement pas très utile pour les développeurs et évaluateurs de systèmes.

La figure 2 résume les différentes classes de fautes qui ont été considérées, classées selon les différents points de vue identifiés. Si toutes les combinaisons de classes de fautes selon les 5 points de vue de la figure 2 étaient possibles, cela conduirait à 32 classes de fautes. En fait, le nombre de combinaisons vraisemblables est plus réduit : 11 combinaisons sont indiquées par les lignes de la figure 3, qui donne également les étiquettes habituellement associées à ces classes de fautes, qu'il ne faut pas confondre avec leur définition ; ces étiquettes sont communément utilisées afin d'exprimer de manière condensée le résultat de la combinaison de plusieurs points de vue.

4.2- Erreurs

Une erreur a été définie comme étant *susceptible* de provoquer une défaillance. Le fait qu'une erreur conduise ou non à défaillance dépend de deux facteurs principaux:

1) La composition du système, et particulièrement la nature de la redondance existante :
   - redondance *intentionnelle* (introduite pour tolérer les fautes) qui est explicitement destinée à éviter qu'une erreur ne conduise à une défaillance,
   - redondance *inintentionnelle* (il est en pratique difficile sinon impossible de construire un système sans aucune forme de redondance[10]) qui peut avoir le même effet – mais inattendu – que la redondance intentionnelle.
2) L'activité du système : une erreur peut être écrasée avant de créer des dommages.
3) La définition d'une défaillance du point de vue de l'utilisateur : ce qui est une défaillance pour un utilisateur donné peut n'être qu'une nuisance supportable pour un autre utilisateur. Des exemples sont a) la prise en compte de la granularité temporelle de l'utilisateur, selon laquelle une erreur qui "traverse" l'interface système-utilisateur peut ou non être considérée comme une défaillance, b) la notion de "taux

[10] Un problème classique pour le test du matériel est précisément l'élimination de telles "fausses redondances", qui peuvent avoir pour effet de masquer certaines fautes, et qui compliquent donc l'élaboration d'entrées de test.

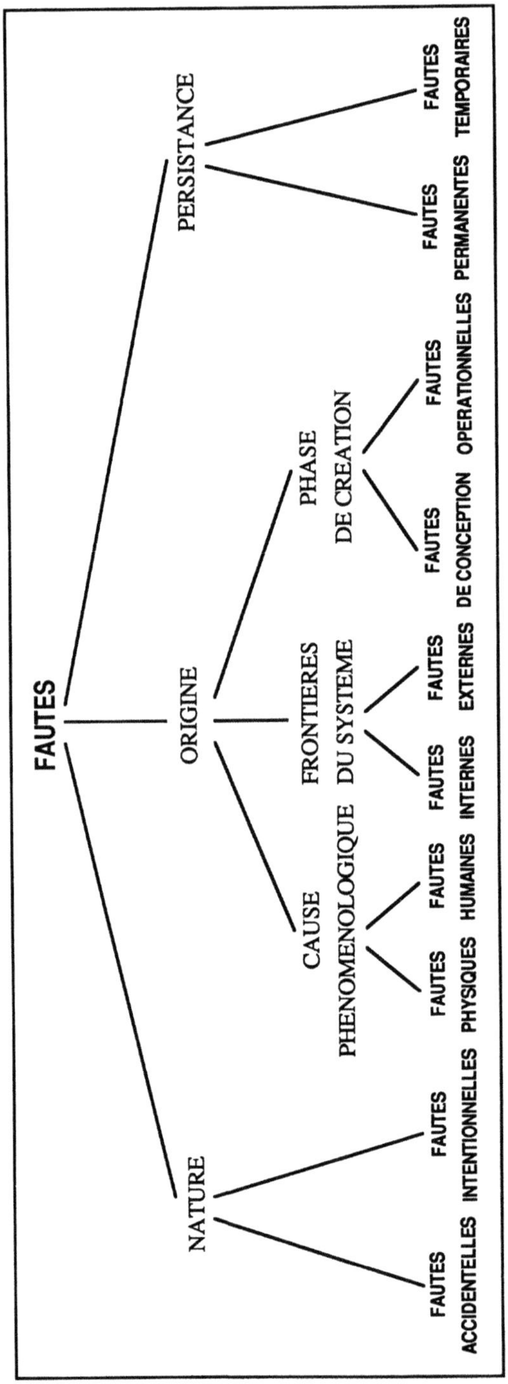

<u>Figure 2</u> - Les classes de fautes selon les différents points de vue

Nature		Origine						Persistance		Appellation usuelle
		Cause phénoménologique		Frontières du système		Phase de création				
Fautes accidentelles	Fautes intentionnelles	Fautes physiques	Fautes humaines	Fautes internes	Fautes externes	Fautes de conception	Fautes opérationnelles	Fautes permanentes	Fautes temporaires	
✓		✓		✓			✓	✓		Fautes physiques
✓		✓			✓		✓	✓		
✓		✓			✓		✓		✓	Fautes transitoires
✓		✓		✓			✓		✓	Fautes intermittentes
✓			✓	✓		✓			✓	
✓			✓	✓		✓		✓		Fautes de conception
✓			✓		✓		✓		✓	Fautes d'interaction
	✓		✓	✓		✓		✓		Logique maligne
	✓		✓	✓		✓			✓	
	✓		✓		✓		✓	✓		Intrusions
	✓		✓		✓		✓		✓	

Figure 3 - Les classes de fautes résultant des combinaisons selon les points de vue de la figure 2 d'erreur acceptable" – implicitement avant de considérer qu'une défaillance est survenue – qui est classique en transmission de données.

Cette discussion illustre pourquoi il est souvent souhaitable de mentionner explicitement dans la spécification des conditions telle que la durée maximum admissible de non-délivrance du service (en fonction de la granularité temporelle de l'utilisateur).

4.3- Défaillances

En fonction de la définition adoptée pour un système d'un point de vue structurel, la notion de défaillance s'applique naturellement tant aux composants qu'aux systèmes, puisqu'un composant est lui-même un système.

Un système ne défaille généralement pas toujours de la même façon, ce qui conduit à la notion de *mode de défaillance*, qui peut être caractérisée selon trois points de vue : domaine de défaillance, perception des défaillances par les utilisateurs du système, conséquences des défaillances sur l'environnement du système.

Le *domaine de défaillance* conduit à distinguer :

- les **défaillances en valeur** : la valeur du service délivré n'est pas conforme à la spécification ;
- les **défaillances temporelles** : les conditions temporelles de délivrance du service ne sont pas conformes à la spécification.

Ces définitions générales (non conformité à la spécification) caractérisent des **défaillances arbitraires**. Il est possible d'affiner les modes de défaillance. Par exemple, on peut distinguer pour les défaillances temporelles les défaillances temporelles *avance* ou *retard*, selon que le service est délivré trop tôt ou trop tard. Une classe de défaillances relative à la fois aux valeurs et aux conditions temporelles est constituée par les **défaillances par arrêt** : l'activité du système, si tant est qu'il en ait une, n'est plus perceptible aux utilisateurs, et une valeur constante du service est délivrée ; cette valeur constante peut varier selon l'application, par exemple la dernière valeur correcte, une valeur prédéterminée, ... Un cas particulier de défaillance par arrêt est constitué par les **défaillances par omission** [Cri 85b, Ezh 86] : aucun service n'est délivré. Une telle défaillance peut être vue comme une limite commune aux défaillances en valeur (pas de valeur) et aux défaillances temporelles (défaillance infiniment tardive) ; une défaillance par omission *persistante* est une **défaillance par écrasement**. Un système dont toutes les défaillances sont — ou plus généralement dont les défaillances sont dans une mesure acceptable — des défaillances par arrêt est un **système à arrêt sur défaillance** ;

un système dont toutes les défaillances sont — dans une mesure acceptable — des défaillances par écrasement est un **système à silence sur défaillance**[11] [Pow 88].

Lorsqu'un système a plusieurs utilisateurs, leurs *perceptions des défaillances* conduit à distinguer :

- les **défaillances cohérentes** : tous les utilisateurs du sytème ont la même perception des défaillances ;
- les **défaillances incohérentes** : les utilisateurs du système peuvent avoir des perceptions différentes des défaillances ; les défaillances incohérentes sont souvent qualifiées de *défaillances byzantines* [Lam 82].

La sévérité des défaillances résulte du classement des *conséquences des défaillances* sur l'environnement du système. Elles permettent donc d'ordonner les modes de défaillance. Un cas particulièrement intéressant est constitué par les systèmes dont les modes de défaillance peuvent être groupés en deux classes dont les sévérités diffèrent considérablement :

- **défaillances bénignes**, dont les conséquences sont du même ordre de grandeur (généralement en termes économiques) que le bénéfice procuré par le service délivré en l'absence de défaillance ;
- **défaillances catastrophiques**, dont les conséquences sont incommensurablement supérieures au bénéfice procuré par le service délivré en l'absence de défaillance.

Un système dont toutes les défaillances sont — dans une mesure acceptable — des défaillances bénignes est un **système sûr en présence de défaillance**. La notion de sévérité des défaillances permet de définir la notion de criticité : la criticité d'un système est la plus forte sévérité de ses modes de défaillance[12]. La relation entre modes de défaillance et criticités des défaillances est conditionnée par l'application considérée. Cependant, il existe une large classe d'applications pour lesquelles l'inactivité est considérée comme une position naturellement sûre (par exemple, transports terrestres, production

---

11 La notion de *processeur* à arrêt sur défaillance telle que définie dans [Sch 83] dans le contexte de systèmes répartis n'est autre qu'un exemple de système à silence sur défaillance.

12 A titre d'exemple, les niveaux de criticité en usage dans l'aviation sont définis comme suit [RTC 85] :
- critique : fonctions pour lesquelles l'occurrence d'une quelconque défaillance empêcherait la déroulement d'un vol ou d'un atterrissage en toute sécurité ;
- essentiel : fonctions pour lesquelles l'occurrence d'une quelconque défaillance réduirait les capacités de l'avion ou l'aptitude de l'équipage à faire face à des conditions opérationnelles contraires ;
- non-essentiel : fonctions dont la défaillance ne pourrait dégrader significativement les capacités de l'avion ou les aptitudes de l'équipage.

d'énergie), d'où la correspondance directe qui existe souvent entre arrêt sur défaillance et sûreté en présence de défaillances [Min 67, Nic 89].

4.4- Pathologie des fautes

Les mécanismes de création et de manifestation des fautes, erreurs, défaillances peuvent être résumés comme suit :

1) Une faute est **active** lorsqu'elle produit une erreur. Une faute active est soit une faute interne qui était préalablement **dormante** et qui a été activée par le processus de traitement, soit une faute externe. Une faute interne peut cycler entre ses états dormant et actif. Les fautes physiques ne peuvent affecter directement que des composants matériels, alors que les fautes humaines peuvent affecter n'importe quel type de composant .
2) Une erreur peut être latente ou détectée ; une erreur est **latente** tant qu'elle n'a pas été reconnue en tant que telle ; une erreur est **détectée** par un algorithme ou un mécanisme de détection. Une erreur peut disparaître avant d'être détectée. Une erreur peut propager[13] — et en général le fait ; en propageant, une erreur crée de nouvelles erreurs.
3) Une défaillance survient lorsqu'une erreur "passe à travers" l'interface système-utilisateur(s) et affecte le service délivré par le système. La conséquence de la défaillance d'un composant est une faute pour le système qui le contient ou pour le, ou les composants qui interagissent avec lui ; les modes de défaillance d'un composant sont donc des types de fautes pour le système ou pour les composants qui interagissent avec lui.

Ces mécanismes permettent de compléter la "chaîne fondamentale" donnée ci-dessous.

... → défaillance → faute → erreur → défaillance → faute → ...

Quelques exemples illustratifs de la pathologie des fautes :

- la conséquence de l'*erreur* d'un programmeur est une *faute* (dormante) dans le logiciel écrit (instruction ou donnée fautive, y compris omission, telle que variable non initialisée) ; lorsque cette faute sera sensibilisée (composant siège de la faute sollicité et données d'entrée telles que l'instruction ou la donnée fautive est activée), elle devient active et produit une *erreur* ; lorsque les données erronées affectent le service

[13] La forme non-réflexive de "propager" est utilisée intentionnellement : une erreur ne se propage pas elle-même.

délivré (en valeur ou en instant de délivrance), une *défaillance* survient ;

- un court-circuit qui se produit dans un circuit intégré est une *défaillance* (par rapport à la spécification du circuit) ; la conséquence est une *faute* (connexion collée à une valeur binaire, modification de la fonction du circuit, etc.), qui restera dormante tant qu'elle ne sera pas activée ; la suite du processus est identique à l'exemple précédent.
- une perturbation électromagnétique d'énergie suffisante est une *faute* ; cette faute peut :
  - directement créer une *erreur* par interférence électromagnétique avec les charges électriques circulant dans les connexions,
  - créer une autre faute, interne ; par exemple, si la perturbation agit sur les entrées d'une mémoire en position d'écriture en changeant les valeurs de certains bits, ces valeurs changées deviendront des fautes lorsque la mémoire ne sera plus sollicitée, qui resteront dormantes jusqu'à ce que la, ou les positions mémoire correspondantes soient lues ; la séquence erreur-défaillance depuis la faute externe jusqu'à la faute interne existe encore, mais au niveau électronique ;
- une interaction homme-machine inappropriée effectuée par un opérateur durant la vie opérationnelle du système est une *faute*, du point de vue du système ; l'altération des données traitées qui en résulte est une *erreur*, etc.
- l'*erreur* d'un rédacteur de manuel de maintenance ou d'opération peut résulter en une *faute* dans le manuel correspondant, sous la forme d'une ou plusieurs directives fautives ; cette faute restera dormante tant que la ou les directives ne sont pas appliquées pour faire face à une situation donnée, etc.

Les exemples ci-dessus montrent clairement que la dormance des fautes peut varier considérablement, en fonction des fautes considérées, de l'utilisation qui est faite du système, etc.

Les fautes humaines peuvent être accidentelles ou intentionnelles. L'exemple ci-dessus relatif à l'erreur d'un programmeur et à ses conséquences peut être rephrasé comme suit : une bombe logique est créée par un programmeur malicieux ; elle restera dormante jusqu'à ce qu'elle soit activée (par exemple, à une date prédéterminée) ; elle produit alors une erreur qui peut conduire à un dépassement de capacité mémoire ou à un ralentissement de l'exécution du programme ; en conséquence, la délivrance du service subira ce qu'on appelle un déni de service, qui est un type particulier de défaillance.

Ces exemples ont été délibérément choisis pour leur simplicité. La réalité est généralement plus complexe. Quatre exemples :

a) une faute donnée dans un composant donné peut résulter de différentes sources possibles ; par exemple, une faute permanente dans un composant matériel – telle que collage à la masse – peut résulter :
   - d'une défaillance physique (par exemple provoquée par une variation de seuil),
   - d'une erreur due à une faute de conception (par exemple une microinstruction fautive) propageant en descendant les couches et causant un court-circuit entre deux sorties du circuit de durée suffisamment longue pour provoquer une défaillance ayant les mêmes effets qu'une variation de seuil ;

b) une faute d'une classe donnée peut, via le phénomène de propagation d'erreur, créer une faute d'une autre classe ; ainsi, dans l'exemple précédent, une faute transitoire pourrait avoir entraîné une exécution erronée de la microinstruction ;

c) certains points de vue peuvent devenir — au moins temporairement — d'une importance moindre durant le processus de propagation ; par exemple, lorsque l'on considère des fautes externes produisant des erreurs d'entrée durant l'exécution d'un composant logiciel (donc l'invoquant dans ce que l'on appelle son domaine d'entrée exceptionnel [Cri 80]), le fait que la faute soit physique ou due à l'homme peut ne pas avoir d'influence significative sur l'occurrence de la défaillance du composant considéré ;

d) une défaillance résulte souvent de l'action combinée de plusieurs fautes ; ceci est particulièrement vrai lorsque l'on s'intéresse à la sécurité-confidentialité : une porte dérobée (c'est-à-dire, un moyen de contourner les contrôles d'accès) qui est insérée dans un système informatique, soit accidentellement, soit intentionnellement, est une faute de conception ; cette faute peut rester dormante jusqu'à ce que par exemple un opérateur malicieux s'en serve pour pénétrer le système en outrepassant ses droits, devenant ainsi un intrus ; le démarrage de session par l'intrus est une faute d'interaction intentionnelle ; durant la session, il peut délibérément créer une erreur, par exemple en modifiant un fichier (attaque de l'intégrité) ; lorsque le fichier correspondant est utilisé par un utilisateur autorisé, le service pourra s'en trouver affecté, et une défaillance survenir.

Deux commentaires finals, relatifs aux mots, ou plutôt aux étiquettes, "faute", "erreur" et "défaillance" :

a) leur utilisation exclusive dans ce document a pour but de montrer que trois concepts sont nécessaires et suffisants pour rendre compte des entraves à la sûreté de fonctionnement, ce qui ne préjuge pas de l'utilisation dans des situations particulières soit de synonymes, soit de mots qui désignent, d'une manière condensée et non ambiguë, une classe spécifique d'entrave ; ceci est tout particulièrement vrai pour les fautes (par exemple bogue, défaut) et pour les défaillances (par exemple, rupture de service, dysfonctionnement) ; il est cependant à noter que le terme "panne", bien que d'un usage courant en Français, est d'une utilisation délicate car il peut désigner tantôt une défaillance (comme dans "tomber en panne"), tantôt une faute (comme dans "trouver la panne") ;
b) l'assignation faite ici pour les étiquettes faute, erreur, défaillance tient simplement compte de l'usage : i) évitement des fautes, tolérance aux fautes, diagnostic de faute, ii) détection et correction d'erreur, iii) taux de défaillance.

Enfin, il importe de souligner que les définitions données dans ce paragraphe sont de nature syntactique ; les critères présidant aux diverses classifications ont en conséquence été mis en relief, et sont à notre avis plus importants que les classes elle-mêmes.

5- Les moyens pour la sureté de fonctionnement

5.1- Dépendances entre les moyens pour la sûreté de fonctionnement

Tous les "comment" qui apparaissent dans les définitions de base du § 1 sont en fait des buts qui ne peuvent être complètement atteints, en raison de la nature humaine, et donc imparfaite, des activités correspondantes. Ces imperfections entraînent des dépendances entre les moyens de la sûreté de fonctionnement, ce qui motive le fait que seule une utilisation combinée — de préférence à chaque étape du processus de conception et de réalisation — de ces moyens peut conduire à un système qui soit sûr de fonctionnement. Ces dépendances peuvent être schématisées comme suit : en dépit de la prévention des fautes grâce à des méthodologies de conception et à des règles de construction (forcément imparfaites afin d'être utilisables), des erreurs surviennent, résultant en des fautes. D'où la nécessité de l'élimination des fautes : lorsqu'une erreur est générée durant la vérification, un diagnostic est entrepris afin de déterminer la, ou les fautes causes de l'erreur, afin de les éliminer. L'élimination des fautes est elle-même imparfaite, de même que les composants pré-existants — matériels ou logiciels — incorporés dans le système, d'où l'importance de la prévision des fautes. Notre dépendance croissante dans les systèmes informatiques conduit à la tolérance aux fautes, qui est à son tour basée sur des règles de construction; d'où à nouveau élimination des fautes, prévision des fautes, ... Il est à noter que le processus est encore plus récursif que ce qui précède le fait apparaître : les systèmes informatiques actuels sont d'une complexité telle que leur développement nécessite des outils informatisés afin d'être efficaces. Ces outils doivent eux-mêmes être sûrs de fonctionnement, et ainsi de suite.

Ce raisonnement illustre les interactions fortes qui existent entre élimination des fautes et prévision des fautes, et fournit donc une motivation à leur association au sein de la *validation*. La validation est souvent limitée à ce qui a été dénommé ici élimination de fautes, auquel cas les deux termes sont souvent associés, comme dans "V et V" [Boe 79], la distinction étant relative à la différence entre "bien construire le système", relié à la vérification, et "construire le bon système", relié à la validation[14]. Ce qui est proposé est simplement une extension de cette vue, la question "le système est-il bon ?" étant complétée par "pour combien de temps le système sera-t-il bon ?"[15]. De plus, l'élimination des fautes est souvent étroitement associée à la prévention des fautes, l'ensemble des deux constituant l'**évitement des fautes**, c'est-à-dire comment *tendre vers* un système sans faute. Outre la mise en évidence de la nécessité de valider la tolérance aux fautes, considérer élimination des fautes et prévision des fautes comme deux constituants de la même activité, la validation, permet une meilleure compréhension de la notion de couverture, et donc d'un important problème introduit par la récursion précédente : *la validation de la validation*, ou comment avoir confiance dans les méthodes et les outils utilisés pour avoir confiance dans le système. La **couverture** se réfère ici à une mesure de la représentativité des situations auxquelles le système est soumis durant sa validation par rapport aux situations auxquelles il sera confronté

---

14 Il est à noter que ces assignations pour vérification et validation sont parfois permutées, comme dans le domaine des protocoles de communication (voir par exemple [Rud 85]).

15 "Validation" vient de "validité", qui englobe deux notions :
- validité à un instant donné, qui est reliée à l'élimination des fautes,
- validité pour une durée donnée, qui est reliée à la prévision des fautes.

durant sa vie opérationnelle[16]. Une couverture imparfaite renforce les relations entre élimination des fautes et prévision des fautes : la nécessité de cette dernière résulte finalement des imperfections de la couverture de l'élimination des fautes.

Dans ce qui suit, nous examinons successivement la tolérance aux fautes, l'élimination des fautes et la prévision des fautes ; nous ne traitons pas de la prévention des fautes car elle relève de l'ingénierie "générale" des systèmes.

5.2- Tolérance aux Fautes

La tolérance aux fautes [Avi 67] est mise en œuvre par le traitement des erreurs et par le traitement des fautes [And 81]. Le **traitement d'erreur** est destiné à éliminer les erreurs, si possible avant qu'une défaillance ne survienne. Le **traitement de faute** est destiné à éviter qu'une, ou des fautes ne soient activées à nouveau.

Le *traitement d'erreur* peut revêtir deux formes :

- **recouvrement d'erreur**, où un état exempt d'erreur est substitué à l'état erroné ; la substitution peut elle-même prendre deux formes [And 81] :
  - **reprise**, où le système est ramené dans un état survenu avant l'occurrence d'erreur ; ceci passe par l'établissement de **points de reprise**, qui sont des instants durant l'exécution d'un processus dont l'état courant peut ultérieurement nécessiter d'être restauré ;
  - **poursuite**, où la transformation de l'état erroné consiste à trouver un nouvel état à partir duquel le système peut fonctionner (habituellement en mode dégradé) ;
- **compensation d'erreur**, où l'état erroné comporte suffisamment de redondance pour permettre la délivrance d'un service non entaché d'erreur à partir de l'état (interne) erroné.

Lorsque le recouvrement d'erreur est utilisé, il est nécessaire que l'état erroné soit identifié comme tel avant de pouvoir le transformer ; ceci est le but de la **détection d'erreur**, d'où la dénomination *détection et recouvrement d'erreur* qui est fréquemment employée. L'adjonction dans un composant de mécanismes de détection d'erreur à ses capacités de traitement fonctionnel

---

[16] La notion de couverture telle que définie ici est très générale ; on peut la préciser en indiquant son champ d'application, par exemple :
- couverture du test d'un logiciel par rapport à son texte, à son graphe de contrôle, ...
- couverture du test d'un circuit intégré par rapport à un modèle de fautes,
- couverture de la tolérance au fautes par rapport à une classe de faute,
- couverture d'une hypothèse de faute par rapport à la réalité.

conduit à la notion de **composant autotestable**, soit matériel [Car 68, Wak 78, Nic 89] soit logiciel [Yau 75, Lap 90a] ; un des principaux intérêts des composants autotestables est la possibilité de définir clairement des *zones de confinement d'erreur* [Sie 82]. Lorsque la compensation d'erreur est effectuée dans un système constitué de composants autotestables partitionnés en classes d'exécution de tâches données, alors la transformation d'état se réduit à la commutation d'un composant défaillant vers un composant non-défaillant au sein de la même classe, d'où la dénomination de *détection et compensation d'erreur* pour l'approche correspondante de la tolérance aux fautes[17]. Au contraire, la compensation d'erreur peut être appliquée systématiquement, même en l'absence d'erreur, procurant alors un **masquage de faute** (par exemple par vote majoritaire). Cependant, ceci peut correspondre à une diminution non perçue de la redondance. C'est pourquoi les mises en œuvre pratiques du masquage comportent généralement de la détection d'erreur, qui peut dans ce cas être effectuée *après* la transformation d'état.

Reprise et poursuite ne sont pas exclusives l'une de l'autre : une reprise peut d'abord être tentée ; si l'erreur persiste, une poursuite peut alors être entreprise. Dans le cas de la poursuite, il est nécessaire d'*estimer les dommages* créés part l'erreur qui a été détectée, ou par les erreurs propagées avant la détection, avant de l'entreprendre ; l'estimation des dommages n'est pas — en principe — nécessaire dans le cas de la reprise, à condition que les mécanismes permettant la transformation de l'état erroné en état exempt d'erreur n'aient pas été affectés [And 81].

Le surcoût temporel (en temps d'exécution) nécessaire pour le traitement d'erreur peut varier considérablement selon la technique adoptée :

- dans le recouvrement d'erreur, le surcoût temporel est plus important lors de l'occurrence d'une erreur qu'en son absence ; dans le cas de la reprise, il consiste à établir des points de reprise, et donc en fait à se préparer au traitement d'erreur ;
- dans la compensation d'erreur, le surcoût temporel est le même, ou pratiquement le même, en présence ou en l'absence d'erreur[18].

---

17 La détection et compensation d'erreur peut être vue comme un cas limite de la détection et recouvrement d'erreur, où le recouvrement est effectué en utilisant l'état présent (erroné) du système au lieu de substituer un état exempt d'erreur à l'état erroné.

18 Dans tous les cas, le temps nécessaire à la mise à jour des tables d'état du système augmente le surcoût temporel.

De plus, la durée d'une compensation d'erreur est beaucoup plus faible que celle d'un recouvrement d'erreur, en raison de la redondance (structurelle) plus importante. Cette remarque

a) est d'une importance pratique certaine, car conditionnant souvent le choix de la stratégie de tolérance aux fautes à adopter par rapport à la granularité temporelle de l'utilisateur du système ;
b) introduit une relation entre redondance temporelle et redondance structurelle ; plus généralement, un système redondant procure toujours un comportement redondant, impliquant un surcoût temporel ; le surcoût temporel peut être suffisamment faible pour ne pas être perçu par l'utilisateur, ce qui signifie seulement que le service n'est pas redondant ; une forme extrême — et opposée — est la "redondance temporelle" (comportement redondant obtenu par répétition) qui nécessite pour être initialisée de la redondance structurelle, limitée mais présente ; on peut donc dire que plus on dispose de redondance structurelle, plus faible sera le surcoût temporel.

La première étape du traitement de faute est le **diagnostic** de faute, qui consiste à déterminer les causes des erreurs, en termes de localisation et de nature. Puis viennent les actions destinées à remplir l'objectif principal du traitement de faute : empêcher une nouvelle activation des fautes, donc la **passivation** des fautes. Ceci est accompli en retirant les composants considérés comme fautifs du processus d'exécution ultérieur. Si le système ne peut plus délivrer le même service qu'auparavant, une *reconfiguration* peut prendre place.

Si l'on estime que le traitement d'erreur a pu directement éliminer la faute, ou si sa probabilité de récurrence est suffisamment faible, l'étape de passivation n'est pas nécessaire. Tant que la passivation des fautes n'est pas entreprise, une faute est considérée comme une **faute douce** ; entreprendre la passivation implique que la faute est considérée comme **dure**, ou **solide**. Au premier abord, les notions de faute douce et de faute dure peuvent sembler synonymes à celles de faute temporaire et de faute permanente. En effet, la tolérance aux fautes temporaires ne nécessite pas de traitement de faute, puisque le recouvrement d'erreur devrait alors éliminer directement les effets de la faute, qui a elle-même disparu, pourvu qu'une faute permanente n'ait pas été créée par le processus de propagation. En fait, les notions de faute douce et de faute dure sont utiles pour les raisons suivantes :

- distinguer une faute permanente d'une faute temporaire est une tâche difficile et complexe, puisqu'une faute temporaire disparaît après une certaine durée, généralement avant que le diagnostic n'ait lieu, et que

des fautes de classes distinctes peuvent donner lieu à des erreurs similaires ; la notion de faute douce ou dure incorpore donc en fait la subjectivité associée à ces difficultés, y compris le fait qu'une faute peut être déclarée douce lorsque le diagnostic est effectué sans succès ;
- leur aptitude à tenir compte de subtilités dans les modes d'action de certaines fautes transitoires ; par exemple, peut-on dire qu'une faute dormante résultant de l'action de particules alpha ou d'ions lourds dans l'espace sur des éléments mémoire (au sens large du terme, y compris des bascules) est une faute *temporaire*? Une telle faute est cependant bien une faute *douce*.

Ce qui vient d'être exposé s'applique tant aux fautes physiques qu'aux fautes de conception : les classes de fautes qui peuvent être réellement tolérées par un système donné dépendent des hypothèses de fautes considérées dans le processus de conception, ce qui est conditionné par l'*indépendance* des redondances par rapport aux processus de création et d'activation des fautes. Un exemple de ce qui précède est donné lorsque l'on considère la tolérance aux fautes physiques et la tolérance aux fautes de conception. Une méthode (largement utilisée) pour la tolérance aux fautes est d'effectuer des traitements multiples par des voies multiples. Lorsque la tolérance aux fautes physiques seules est recherchée, les voies multiples peuvent être identiques, en vertu de l'hypothèse selon laquelle des composants matériels défaillent indépendamment. Une telle approche n'est de toute évidence pas adéquate pour la tolérance aux fautes de conception où les multiples voies doivent délivrer *des services identiques via des conceptions et des réalisations séparées* [Elm 72, Ran 75, Che 78], ce qui est la définition de la **diversification fonctionnelle**.

Dans les systèmes tolérant les fautes, on rencontre fréquemment des situations de fautes et/ou de défaillances multiples. Sous l'angle de leur causes, on est conduit à distinguer :
- les **fautes indépendantes**, qui sont attribuées à des causes différentes ;
- les **fautes corrélées**, qui sont attribuées à une cause commune.

Une autre notion utile est celle d'**erreurs coïncidentes**, c'est-à-dire qui sont créées sur la même entrée [Avi 86].

La relation temporelle entre défaillances multiples conduit à distinguer :
- les **défaillances simultanées**, qui surviennent dans une fenêtre temporelle prédéfinie ;
- les **défaillances séquentielles**, qui ne surviennent pas dans la même fenêtre temporelle prédéfinie.

Bien qu'une définition mathématique de "simultanée" conduirait à une fenêtre de durée nulle, la fenêtre peut être dans la pratique d'une durée significative, dépendante de l'application considérée, d'où la notion de défaillances "presque coïncidentes" [Tri 84]. Typiquement, dans un système tolérant les fautes qui a été conçu pour tolérer une faute à la fois, il est nécessaire de se remettre des effets d'une faute avant que le système soit à même de tolérer une autre faute. Dans ce contexte, la durée qui permet de distinguer défaillances simultanées de défaillances séquentielles est l'intervalle de temps nécessaire pour le traitement d'erreur et éventuellement le traitement de faute, intervalle durant lequel le système est vulnérable.

Un aspect important dans la coordination des activités de composants multiples est d'éviter que la propagation d'erreurs n'affecte l'activité de composants non-défaillants. Cet aspect devient particulièrement important lorsqu'un composant donné doit communiquer à d'autres composants une information qui lui est propre. Des exemples typiques de telles *informations issues de sources uniques* sont des données locales de capteurs, la valeur d'une horloge locale, la perception locale de l'état d'autres composants,... La conséquence de la nécessité de communiquer d'un composant à d'autres composants des informations issues d'une source unique est que les composants non-défaillants doivent se mettre *d'accord* sur la manière d'utiliser de manière cohérente les informations obtenues (voir par exemple les diffusions atomiques [Cri 85a], la synchronisation d'horloges [Lam 85, Kop 87], ou les protocoles d'appartenance [Cri 88]). Il est cependant important de réaliser que l'inévitable présence de redondances structurelles dans tout système tolérant les fautes implique une répartition des ressources à un niveau ou à un autre, et que le problème de l'accord persiste par conséquent. Des systèmes tolérant les fautes géographiquement localisés peuvent apporter au problème de l'accord des solutions qui seraient jugées trop coûteuses dans des systèmes répartis "classiques", communiquant par messages (par exemple, les étages intermédiaires [Lal 86] ou les étages multiples [Fri 82] pour assurer la cohérence interactive).

La connaissance de certaines propriétés du système peut permettre de limiter la redondance nécessaire. Des exemples — classiques — sont fournis par les régularités de nature structurelle : codes détecteurs et correcteurs d'erreurs [Pet 72], structures de données robustes [Tay 80], multiprocesseurs et réseaux [Pra 86, Ren 86], tolérance aux fautes basée sur des algorithmes [Hua 82]. Les fautes tolérées sont alors dépendantes des propriétés considérées puisque celles-ci interviennent directement dans les hypothèses de fautes prises en compte dans le processus de conception.

Signaler la défaillance d'un composant à ses utilisateurs est d'une importance particulière. Ceci peut être pris en compte dans le cadre des exceptions [Mel 77, Cri 80, And 81]. Les primitives de *traitement d'exception* fournies par certains langages peuvent être mises à profit pour la mise en œuvre du traitement d'erreur, en particulier de la poursuite[19].

La tolérance aux fautes est (également) un concept récursif : les mécanismes destinés à mettre en œuvre la tolérance aux fautes doivent être protégés contre les fautes susceptibles de les affecter eux-mêmes. Des exemples sont fournis par la replication de voteurs, par les contrôleurs auto-testables [Car 68], par la notion de mémoire "stable" [Lam 81] pour les programmes et données de recouvrement.

La tolérance aux fautes n'est pas limitée aux fautes accidentelles. La protection contre des intrusions fait traditionnellement appel à la cryptographie [Den 82]. Certains mécanismes de détection d'erreur sont destinés tant aux fautes accidentelles qu'aux fautes intentionnelles (par exemple, les techniques de protection d'accès aux mémoires), et des approches ont été proposées pour tolérer à la fois les intrusions et les fautes physiques [Fra 86, Rab 89], ainsi que pour tolérer des logiques malignes [Jos 88].

5.3- Elimination des fautes

L'élimination des fautes est constituée de trois étapes : vérification, diagnostic, correction. La **vérification** consiste à déterminer si le système satisfait des propriétés, appelées *conditions de vérification* [Che 81] ; si ce n'est pas le cas, les deux autres étapes doivent être entreprises : diagnostiquer la ou les fautes qui ont empéché les conditions de vérification d'être remplies, puis apporter les corrections nécessaires. Après correction, le processus doit être recommencé afin de s'assurer que l'élimination de faute n'a pas eu de conséquences indésirables ; les vérifications ainsi effectuées sont généralement qualifiées de **non-régression**. Les vérifications peuvent prendre deux formes :

- conditions générales, qui s'appliquent à une classe donnée de système, et sont par conséquent — relativement — indépendantes des spécifications, par exemple absence de verrou mortel, conformité à des règles de conception et de réalisation ;

---

[19] L'utilisation du terme "exception", due à ses origines de confrontation à des situations exceptionnelles — pas seulement des erreurs — est à utiliser avec précaution dans le contexte de la tolérance aux fautes : il pourrait apparaître comme contradictoire avec le fait de considérer la tolérance aux fautes comme un attribut naturel des systèmes informatiques, prise en compte depuis les toutes premières phases de la conception, et non comme un attribut "exceptionnel".

- conditions spécifiques du système considéré, directement déduites des spécifications.

Les techniques de vérification peuvent être classées selon qu'elles impliquent ou non l'activation du système. La vérification d'un système sans activation réelle est la **vérification statique**, qui peut être conduite :

- sur le système lui-même, sous la forme soit d'une *analyse statique* (par exemple inspections [Mye 79], analyse du flot de données [Ost 76], analyse de complexité [McC 76], vérifications effectuées par les compilateurs, ...), soit d'une *preuve mathématique* (assertions inductives [Hoa 69, Cra 87]) ;
- sur un modèle de comportement du système (par exemple réseaux de Petri, automates à états finis), conduisant à une analyse de comportement [Dia 82].

Vérifier un système en l'activant constitue la **vérification dynamique** : les entrées fournies au système peuvent être symboliques dans le cas de l'**exécution symbolique**, ou valuées dans le cas du test de vérification, habituellement appelé simplement **test**.

Le test exhaustif d'un système par rapport à toutes ses entrées possibles est pratiquement infaisable. Les méthodes de détermination des jeux de test peuvent être classées selon deux points de vue : critères pour sélectionner les entrées de test, génération des entrées de test.

Les *critères* de sélection des entrées de test peuvent à leur tour être classés selon trois points de vue :

- but du test : vérifier si un système satisfait ses spécifications est le **test de conformité**, alors qu'un test destiné à révéler des fautes est appelé **test de recherche de fautes** ;
- modèle du système : selon que le modéle est relatif à la fonction ou à la structure du système, on est conduit à du **test fonctionnel** ou à du **test structurel** ;
- l'existence d'un modèle de faute : l'existence d'un tel modèle conduit au **test basé sur des fautes** [Mor 90], destiné à révéler des classes de fautes spécifiques (par exemple, fautes de collage issues de la production du matériel [Rot 67], fautes physiques affectant le répertoire d'instructions d'un microproceseur [Tha 78], fautes de conception du logiciel [Goo 75, DeM 78, How 87]) ; en l'absence de modèle de faute, les critères de sélection peuvent être relatifs à la sensibilisation de

chemins [Rap 85, Nta 88], aux valeurs limites des entrées [Mye 79] dans le logiciel, ...[20]

La *génération* des entrées de test peut être déterministe ou probabiliste :

- dans le **test déterministe**, les jeux de test sont déterminés par un choix sélectif selon le critère retenu :;
- dans le **test aléatoire**, également appelé **test statistique**, les jeux de test sont sélectionnés selon une distribution probabiliste du domaine d'entrée, la distribution et le nombre de données d'entrée étant déterminés selon le critère retenu [Dav 81, Dur 84].

La combinaison des différents points de vue identifiés conduit aux différentes approches de test, certaines pouvant recevoir une dénomination condensée ; à titre d'exemples :

- le test structurel du matériel signifie généralement test de recherche de fautes, structurel, basé sur des fautes, alors qu'il signifie test de recherche de fautes, structurel, non basé sur des fautes lorsque l'on considère le logiciel ;
- le *test de mutation* du logiciel [DeM 78] est un test de recherche de fautes, structurel, basé sur des fautes, déterministe.

Observer les sorties de test et décider si elles satisfont les conditions de vérification est généralement connu sous le nom de *problème de l'oracle* [Adr 82]. Les conditions de vérification peuvent s'appliquer à l'ensemble des données de sortie ou à une compression de ces dernières (par exemple, signature dans le test de fautes physiques du matériel [Dav 86], ou "oracle partiel" dans le test du logiciel [Wey 82]). Lorsque le test est relatif à des fautes physiques, les résultats — compressés ou non — attendus du système sous test pour une séquence d'entrées donnée peuvent être déterminés par simulation [Lev 86] ou à partir d'un système de référence. Pour les fautes de conception, la référence est généralement la spécification ; elle peut également être constituée par un prototype, ou par une autre mise en œuvre de la même spécification dans le cas de conception diversifiée (on parle alors de "test dos-à-dos", voir par exemple [Bis 88]).

Certaines méthodes de vérification peuvent être utilisées conjointement, par exemple l'exécution symbolique peut être utilisée pour déterminer des jeux d'entrée [Adr 82], ou en tant que méthode de preuve mathématique [Car 78].

---

[20] La possibilité de définir un modèle de faute est étroitement liée au stade de développement considéré : la possibilité augmente avec l'avancement.

Des vérifications devant être effectuées tout au long du développement d'un système, les techniques qui viennent d'être évoquées s'appliquent naturellement aux diverses formes que revêt un système au cours de son développement : prototype, composant, ...

Vérifier qu'un système ne peut faire plus que ce qui est spécifié est tout particulièrement important vis-à-vis des fautes intentionnelles [Gas 88].

La conception d'un système afin de faciliter sa vérification est la **conception pour la vérification.** Cette approche est particulièrement développée pour le matériel vis-à-vis des fautes physiques, où les techniques correspondantes sont appelées *conception en vue du test* [Wil 83, McC 86].

L'élimination de fautes durant la vie opérationnelle d'un système est la **maintenance corrective**, destinée à préserver ou à améliorer l'aptitude du système à délivrer un service en conformité avec la spécification[21]. La maintenance corrective peut revêtir deux formes :

- **maintenance curative**, destinée à éliminer des fautes ayant produit des erreurs qui ont été signalées ;
- **maintenance préventive**, destinée à éliminer des fautes avant qu'elles ne produisent des erreurs ; ces fautes peuvent être :
  - des fautes physiques survenues dans le système considéré depuis les dernières actions de maintenance préventive,
  - des fautes de conception ayant provoqué des erreurs dans d'autres systèmes similaires [Ada 84].

Ces définitions[22] s'appliquent tant aux systèmes non tolérants aux fautes qu'aux systèmes tolérants aux fautes ; ces derniers peuvent être maintenables en ligne (sans interrompre la délivrance du service) ou hors ligne. Il est enfin à noter que la frontière entre maintenance corrective et traitement de faute est

---

21 Les autres formes de maintenance habituellement considérées sont [Ram 84] :
- la maintenance adaptative, qui a pour but d'adapter le système à des modifications de son environnement (par exemple, changement de système d'exploitation ou de système de gestion de bases de données) ;
- maintenance perfective, qui a pour but d'améliorer les fonctions du système, en réponse aux souhaits des utilisateurs — ou des concepteurs, et qui peut comporter l'élimination de fautes dans les spécifications.

22 Il est à noter que les discussions actuelles sur le bien-fondé de l'emploi du terme maintenance lorsqu'appliqué au logiciel, et plus particulièrement pour les formes de maintenance dites "adaptative" et "perfective" oublient simplement l'étymologie : au Moyen Age, maintenance désignait les actions nécessaires pour maintenir une armée dans l'état de livrer bataille, donc incluant les deux formes précédemment mentionnées. L'association de maintenance avec réparation de matériel est donc en fait une déviation – récente. Associer "maintenir" à la notion de service permettrait de faire revivre cette signification étymologique, et éliminerait la source même de discussion.

relativement arbitraire ; en particulier, la maintenance curative peut être considérée comme un moyen — ultime — de tolérance aux fautes.

5.4- Prévision des fautes

La prévision des fautes est conduite en effectuant des évaluations du comportement du système par rapport à l'occurrence des fautes et à leur activation. L'évaluation a deux facettes :

- non-probabiliste, par exemple via la détermination des coupes ou chemins minimaux dans un arbre de fautes, ou la conduite d'une analyse de modes de défaillance ;
- probabiliste, destinée à établir la conformité du système à des objectifs de sûreté de fonctionnement établis en termes de probabilités associées à certains attributs de la sûreté de fonctionnement, qui peuvent alors être vus comme des mesures de la sûreté de fonctionnement.

La vie d'un système est perçue par son, ou ses utilisateurs comme une alternance entre deux états du service par rapport à la spécification :

- **service correct**, où le service délivré *est conforme* à la spécification[23] ;
- **service incorrect**, où le service délivré *n'est pas conforme* à la spécification·

Une défaillance est donc une transition de service correct à service incorrect, et la transition de service incorrect à service correct est une **restauration**. La quantification de l'alternance entre service correct et service incorrect permet de définir fiabilité et disponibilité comme des *mesures* de la sûreté de fonctionnement :

- **fiabilité** : mesure de la délivrance *continue* d'un service correct, ou, de façon équivalente, du temps *jusqu'à* défaillance ;
- **disponibilité** : mesure de la délivrance d'un service correct *par rapport à l'alternance* service correct-service incorrect.

Une troisième mesure, la **maintenabilité**, est habituellement considérée, qui est définie comme la mesure du temps jusqu'à restauration depuis la dernière défaillance survenue, ou, ce qui est équivalent, de la délivrance continue d'un service incorrect. Cette mesure n'est pas moins importante que la fiabilité et la disponibilité ; elle n'a pas été introduite dans les définitions de

---

23 Nous restreignons délibérément l'utilisation de "correct" au service délivré par le système, et ne l'utilisons pas pour le système lui-même : de notre point de vue, il n'existe guère de système sans faute, tout au plus existe-t-il des systèmes qui n'ont pas encore défailli.

base car elle peut, au moins conceptuellement, être déduite de la connaissance de ces deux dernières.

En tant que mesure, la sécurité-innocuité peut être vue comme une extension de la fiabilité. Groupons les états de service correct et de service incorrect consécutif à une défaillance bénigne en un unique état sûr (au sens de l'absence de dommage catastrophique, non forcément de danger). La **sécurité-innocuité** est alors une mesure du temps jusqu'à défaillance catastrophique. Une extension directe de la disponibilité, c'est-à-dire une mesure exprimant le fait d'être dans l'état sûr par rapport à l'alternance état sûr-état consécutif à une défaillance catastrophique ne serait pas très significative. En effet, lorsqu'une défaillance catastrophique est survenue, les conséquences sont généralement d'une ampleur telle que la restauration du service n'est pas ce qui prime le plus, pour — au moins — les deux raisons suivantes :

- elle devient secondaire par rapport à la réparation (au sens large du terme, y compris juridique) des conséquences de la catastrophe ;
- la longue période avant que le système ne puisse être à nouveau opérationnel (commissions d'enquête, ...) conduirait à des valeurs numériques non significatives.

Une mesure "hybride" du type fiabilité-disponibilité peut cependant être définie : mesure de la délivrance d'un service correct par rapport à l'alternance service correct-service incorrect suite à défaillance bénigne. Cette mesure présente l'intérêt de permettre une quantification de la disponibilité du système avant qu'une défaillance catastrophique ne survienne, et permet par là-même de quantifier l'habituel compromis entre fiabilité (ou disponibilité) et sécurité-innocuité.

Dans le cas de systèmes multi-performants, plusieurs services peuvent être distingués ; de même, on peut distinguer plusieurs modes de délivrance des services depuis la pleine capacité jusqu'à l'interruption totale, ce qui peut être vu comme des délivrances de moins en moins correctes. Les mesures combinées peformances-sûreté de fonctionnement sont habituellement dénommées **performabilité** [Mey 78, Smi 88].

Les deux approches principales de la prévision probabiliste des fautes, destinées à obtenir des estimateurs quantifiés des mesures de la sûreté de fonctionnement, sont la modélisation et le test (d'évaluation). Ces approches sont directement complémentaires, en ce sens que la modélisation nécessite des données relatives aux processus élémentaires modélisés (processus de défaillance, de maintenance, d'activation du système, ...), qui peuvent être obtenues par le test.

Lorsque l'on effectue une évaluation par modélisation, les méthodes diffèrent significativement selon que le système est considéré comme étant en fiabilité stabilisée ou en croissance de fiabilité, qui peuvent être définies comme suit [Lap 90] :

- **fiabilité stabilisée** : l'aptitude du système à délivrer un service correct est *préservée* (identité stochastique des temps jusqu'à défaillance successifs) ;
- **croissance de fiabilité** : l'aptitude du système à délivrer un service correct est *améliorée* (augmentation stochastique des temps jusqu'à défaillance successifs)[24].

Des interprétations pratiques de fiabilité stabilisée et de croissance de fiabilité sont les suivantes :

- fiabilité stabilisée : à une restauration donnée, le système est identique à ce qu'il était à la précédente restauration ; ceci correspond aux situations suivantes :
  - dans le cas d'une défaillance du matériel, le composant défaillant est remplacé par un autre, identique et non-défaillant,
  - dans le cas d'une défaillance du logiciel, le système est relancé sur un point d'entrée différent de celui ayant conduit à défaillance ;
- croissance de fiabilité : la faute dont l'activation a conduit à défaillance est diagnostiquée comme une faute de conception (matérielle ou logicielle) et est éliminée.

L'évaluation de la sûreté de fonctionnement de système en fiabilité stabilisée est généralement composée de deux étapes principales :

- *construction du modèle* du système à partir des processus stochastiques élémentaires qui modélisent le comportement des composants du système et leurs interactions ;
- *traitement du modèle* afin d'obtenir les expressions et les valeurs des mesures de la sûreté de fonctionnement du système.

L'évaluation peut être effectuée par rapport aux fautes physiques [Tri 84], par rapport aux fautes de conception [Lit 79, Arl 88], ou par rapport aux deux [Lap 84, Pig 88]. La sûreté de fonctionnement d'un système est étroitement

---

[24] Une décroissance de fiabilité (l'aptitude du système à délivrer un service correct est *dégradée*, et il y a une diminution stochastique des temps jusqu'à défaillance successifs) est théoriquement, et pratiquement possible ; par exemple, suite à l'introduction de nouvelles fautes durant des actions correctives, fautes dont la probabilité d'activation est supérieure à celle de la, ou des fautes éliminées. Dans une telle situation, il est à souhaiter que la décroissance soit limitée dans le temps, et que la fiabilité soit globalement croissante sur une longue période d'observation.

dépendante de son environnement, que ce soit au sens large du terme [Hec 87], ou plus spécifiquement de sa charge [Cas 81, Iye 82].

De nombreux modèles de croissance de fiabilité ont été proposés, pour le matériel [Dua 64], pour le logiciel, ou pour l'ensemble des deux [Lap 90]. La plupart de ces modèles sont dévolus au logiciel, et sont destinés à l'évaluation soit de la fiabilité [Yam 85, Mil 86], soit du nombre (résiduel) de fautes [Goe 79, Toh 89] ; ces modèles permettant de faire des estimations de fiabilité dans le futur à partir de données accumulées relatives au passé, une attention particulière a été portée aux problèmes de prévision [Lit 88].

Bien que non a priori destinés à vérifier un système, les tests d'évaluation [Mil 87, Cho 87] peuvent être caractérisés en utilisant les points de vue définis au § 5.3 : il s'agit de tests de conformance, fonctionnels, non basés sur des fautes, statistiques. Un souci primordial dans ces tests est la représentativité des conditions opérationnelles par les jeux de test, d'où leur nom : **tests opérationnels**.

Lorsque l'on évalue un système tolérant les fautes, la couverture des mécanismes de traitement des erreurs et des fautes a une influence primordiale [Bou 69, Arn 73] ; son évaluation peut être effectuée par modélisation [Dug 89] ou par test, appelé dans ce cas *injection de fautes* [Arl 89, Gun 89].

6- Les Attributs de la Sureté de Fonctionnement

Les attributs de la sûreté de fonctionnement ont été définis au § 1 selon diverses propriétés, sur lesquelles on peut mettre un accent plus ou moins prononcé selon l'application à laquelle est destiné le système informatique considéré :

- la disponibilité est toujours requise, bien qu'à des degrés naturellement variables selon les applications ;
- fiabilité, sécurité-innocuité, sécurité-confidentialité peuvent être ou ne pas être requises selon les applications.

Une propriété additionnelle, qui peut être vue comme prérequise pour les autres propriétés, est l'**intégrité**, c'est-à-dire le fait de demeurer intact, et donc d'éviter modifications ou suppressions indésirées.

Les variations dans l'accent mis sur les attributs de la sûreté de fonctionnement ont une influence directe sur le dosage approprié des techniques décrites aux paragraphes précédents qu'il convient de déterminer pour que le système résultant soit sûr de fonctionnement. Ceci est un problème d'autant plus délicat que certains attributs sont antagonistes (par exemple, disponibilité et sécurité-innocuité, disponibilité et sécurité-confidentialité), d'où la nécessité de faire des compromis. Lorsque l'on considère les trois principales dimensions de la conception d'un système informatique, c'est-à-dire le coût, les performances et la sûreté de fonctionnement, le problème est rendu encore plus délicat par le fait que la dimension de sûreté de fonctionnement est moins bien maîtrisé que les deux autres dimensions [Sie 82].

La maintenabilité a été définie au § 5.4 comme une mesure de la sûreté de fonctionnement. La **maintenabilité** peut également être considérée en tant qu'attribut de la sûreté de fonctionnement, relatif à la facilité avec laquelle la maintenance d'un système peut être effectuée.

La définition de la sécurité-confidentialité donnée au § 1, la prévention d'accès ou de manipulations non autorisées de l'information, vient de [ITS 90] ; cette référence définit la sécurité-confidentialité comme la combinaison de la confidentialité, prévention de divulgation non autorisée de l'information, de l'intégrité, prévention de modification ou suppression non autorisée de l'information, de la disponibilité, prévention de rétention non autorisée de l'information. Il est à noter que :

a) un accès ou une manipulation non autorisé de l'information peut résulter de fautes accidentelles ou intentionnelles, et que certains mécanismes de protection contre des accès illégitimes sont communs aux deux types de fautes — comme déjà mentionné au § 5.2 ;
b) la notion d'autorisation doit être comprise de façon non restrictive par rapport aux fautes intentionnelles : une personne autorisée qui abuse de son autorisation en fait viole le niveau d'autorisation qui lui avait été accordé en effectuant des actions illégitimes, et devient de ce fait un intrus.

Une caractéristique importante des systèmes tolérant les fautes est leur aptitude, grâce à la présence de procédures de détection d'erreurs, à fournir à leurs utilisateurs des informations indiquant si le service (fonctionnel) qui est délivré est correct ou non. Une telle propriété est appelé **crédibilité** dans [Fur 90].

Déterminer si un système est réellement sûr de fonctionnement — justification de la confiance accordée au service délivré — ou non déborde largement le champ des techniques dont il a été fait état aux paragraphes précédents, et ce pour, au moins, les trois raisons et limites suivantes :

- déterminer avec certitude la couverture des hypothèses de conception ou de validation (par exemple, pertinence des critères utilisés pour déterminer des jeux de test par rapport aux fautes susceptibles de se produire dans la réalité, hypothèses de fautes dans la conception de mécanismes de tolérance aux fautes) impliquerait une connaissance et une maîtrise des technologies utilisées, de l'utilisation envisagée du système, ... qui vont au-delà du réalisable ;
- évaluer un système selon certains attributs de la sûreté de fonctionnement par rapport à certaines classes de fautes est actuellement considéré comme infaisable ou comme ne pouvant procurer des résultats significatifs : les bases théoriques probabilistes n'existent pas, ou ne sont pas — encore — largement acceptées ; par exemple, la sécurité-innocuité par rapport aux fautes de conception, la sécurité-confidentialité par rapport aux fautes intentionnelles ;
- les spécifications par rapport auxquelles la validation est conduite ne sont généralement pas exemptes de fautes — comme tout système.

Parmi les nombreuses conséquences de cet état de choses, on peut mentionner :

- l'accent mis sur le processus de développement et de production : méthodes et techniques employées et comment elles sont employées ; dans certains cas, une note est assignée au système en fonction de la nature des méthodes et techniques employées, ainsi que d'une évaluation de leur emploi[25] ;
- la présence, dans les spécifications de certains systèmes tolérant les fautes, du nombre de fautes que le système doit tolérer, en plus d'objectifs probabilistes en termes de mesures de la sûreté de fonctionnement[26] ; de telles spécifications ne seraient pas nécessaires si les limites mentionnées ci-dessus pouvaient être repoussées.

---

[25] Par exemple :
- du point de vue de la sécurité-confidentialité, les systèmes vont de "A1" ("conception vérifiée") à "D" ("protection minime") [DoD 85] ;
- les logiciels pour les avions civils de transport sont classés [RTC 85] en niveaux 1, 2 ou 3 selon la criticité des fonctions qu'ils accomplissent : critiques, essentiels, non-essentiels.

[26] De telles spécifications sont classiques dans les applications aérospatiales, sous la forme de concaténation d'exigences de la forme "opérationnel sur défaillance" ou "sûr en présence de défaillances".

CONCLUSION

De façon croissante, les personnes et les organisations développent ou s'équipent de systèmes informatiques élaborés dans les services desquels ils doivent placer une confiance grandissante — qu'il s'agisse d'alimenter un réseau de distributeurs de monnaie, de calculer l'orbite d'un satellite, de commander et contrôler un avion ou une centrale nucléaire, ou de maintenir la confidentialité d'une base de données sensibles. Selon les circonstances, l'accent pourra être mis sur différentes propriétés de ces services — par exemple, le temps moyen de réponse obtenu, la vraisemblance des résultats produits, l'aptitude à éviter des défaillances qui pourraient être catastrophiques pour l'environnement du système, ou la mesure dans laquelle des intrusions délibérées peuvent être prévenues. La notion de sûreté de fonctionnement fournit un moyen approprié pour rassembler ces préoccupations diverses au sein d'un cadre conceptuel unique. La sûreté de fonctionnement inclut donc en tant que cas particuliers les propriétés telles que fiabilité, disponibilité, sécurité-innocuité, sécurité-confidentialité. Elle fournit également le moyen d'aborder le problème du *dosage approprié* de ces propriétés afin de satisfaire au mieux les besoins des utilisateurs.

GLOSSAIRE

Ce glossaire récapitule les définitions données tout au long du texte. Il est fourni en tant qu'aide à la lecture, et ne doit pas être considéré isolément.

Attributs de la sûreté de fonctionnement......................Attributs permettant a) d'exprimer les propriétés qui sont attendues du système, et b) d'apprécier la qualité du service délivré, telle que résultant des entraves et des moyens de s'y opposer.

Fiabilité, maintenabilité, disponibilité, sécurité-innocuité, sécurité-confidentialité.

Compensation (d'erreur) .... Forme de traitement d'erreur où l'état erroné comporte suffisamment de redondance pour permettre la délivrance d'un service non entaché d'erreur.

Comportement (d'un système) ... Ce que fait un système.

Composant (d'un système) .... Un autre système.

Composant autotestable .... Composant comportant des mécanismes de détection d'erreur.

Conception pour la vérification .... Méthodes et techniques de conception d'un système destinées à faciliter sa vérification.

Couverture .... Mesure de la représentativité des situations auxquelles le système est soumis durant sa validation par rapport aux situations auxquelles il sera confronté durant sa vie opérationnelle.

Crédibilité .... Aptitude d'un système à fournir à ses utilisateurs des informations indiquant si le service délivré est correct.

Criticité (d'un système) .... Sévérité maximum de ses modes de défaillance.

Croissance de fiabilité .... L'aptitude du système à délivrer un service correct est améliorée (augmentation stochastique des temps jusqu'à défaillance successifs).

Défaillance .... Evénement survenant lorsque le service délivré n'est plus conforme à la spécification.

Transition de service correct vers service incorrect.

Défaillance arbitraire .... voir Défaillance.

Défaillance bénigne .... Défaillance dont les conséquences sont du même ordre de grandeur (généralement en termes économiques) que le bénéfice procuré par le service délivré en l'absence de défaillance.

Défaillance catastrophique .... Défaillance dont les conséquences sont incommensurablement supérieures au bénéfice retiré de la délivrance d'un service en l'absence de défaillance.

Défaillance cohérente .... Défaillance perçue identiquement par tous les utilisateurs du système.

Défaillances de mode commun ..Défaillances résultant de fautes corrélées.

Défaillance en valeur ....Défaillance telle que la valeur du service délivré n'est pas conforme à la spécification.

Défaillance incohérente ....Défaillance dont les utilisateurs du système peuvent avoir des perceptions différentes.

Défaillance par arrêt..................L'activité du système, si tant est qu'il en ait une, n'est plus perceptible aux utilisateurs, et une valeur constante du service est délivrée.

Défaillance par écrasement........Défaillance par omission persistante.

Défaillance par omission ....Aucun service n'est délivré.

Défaillances séquentielles ....Défaillances qui ne surviennent pas dans la même fenêtre temporelle prédéfinie.

Défaillances simultanées ....Défaillances qui surviennent dans une fenêtre temporelle prédéfinie.

Défaillance temporelle...............Défaillance telle que les conditions temporelles de délivrance du service ne sont pas conformes à la spécification.

Détection (d'erreur)...................Identification d'un état erroné comme tel.

Diagnostic (de faute)..................Détermination des causes des erreurs, en termes de localisation et de nature.

Disponibilité ....Sûreté de fonctionnement par rapport au fait d'être prêt à l'utilisation.

Mesure de la délivrance d'un service correct par rapport à l'alternance service correct-service incorrect.

Diversification fonctionnelle.....Approche pour le développement d'un système destinée à fournir des services identiques via des conceptions et des réalisations séparées.

Elimination des fautes...............Méthodes et techniques destinées à réduire la présence (en nombre et en sévérité) des fautes.

Entraves (à la sûreté de fonctionnement).........................Circonstances indésirables, mais non inattendues, causes ou résultats de la non-sûreté de fonctionnement

Fautes, erreurs, défaillances.

Environnement (d'un système)..Les autres systèmes ayant interagi ou interféré, interagissant ou interférant, ou susceptible d'interagir ou d'interférer avec le système considéré.

Erreur..Partie de l'état d'un système – par rapport au processus de traitement – qui est susceptible d'entraîner une défaillance.

Manifestation d'une faute dans un système

Erreurs coïncidentes..................Erreurs produites sur la même entrée.

Erreur détectée..........................Erreur reconnue en tant que telle par un algorithme ou un mécanisme de détection.

Erreur latente.............................Erreur qui n'a pas été reconnue en tant que telle.

Etat (d'un système)....................Condition d'être par rapport à un ensemble de circonstances.

Evitement des fautes..................Méthodes et techniques permettant de tendre vers un système exempt de fautes.

Prévention des fautes et élimination des fautes.

Exécution symbolique................Vérification dynamique effectuée avec des entrées sous forme de symboles.

Faute...Cause adjugée ou supposée d'une erreur.

Cause d'erreur évitée ou tolérée.

Conséquence de la défaillance d'un composant pour le système qui le contient ou pour le, ou les composants qui interagissent avec lui.

Faute accidentelle ....Faute apparaissant ou créée de manière fortuite.

Faute active ....Faute qui produit une erreur.

Fautes corrélées..........................Fautes attribuées à une cause commune.

Faute de conception ....Résultat d'imperfections commises soit au cours du développement d'un système (de l'expression des besoins à la recette, y compris l'établissement des procédures d'exploitation ou de maintenance), soit au cours de modifications ultérieures.

Faute dormante...........................Faute interne non activée par le processus de traitement.

Faute douce................................Faute éliminée directement par le traitement d'erreur.

Faute dure, ou **solide**..................Faute nécessitant d'être passivée.

Faute externe............................Faute résultant de l'interférence ou des interactions d'un système avec son environnement physique ou humain.

Faute humaine............................Conséquence d'imperfections humaines.

Fautes indépendantes..................Fautes attribuées à des causes différentes.

Faute intentionnelle....................Faute résultant d'une action délibérée.

Faute intermittente......................Faute temporaire interne.

Faute dont les conditions d'activation ne peuvent être reproduites ou qui se manifeste suffisamment rarement.

Faute interne...............................Partie de l'état d'un système qui, lorsqu'activée par les traitements, produira une ou des erreurs.

Faute opérationnelle...................Faute apparaissant durant l'exploitation du système.

Faute permanente........................Faute dont la présence n'est pas reliée à des conditions ponctuelles, internes (processus de traitement) ou externes (environnement).

Faute physique............................Conséquence de phénomènes physiques adverses.

Faute temporaire........................Faute qui n'est présente que pour une durée limitée.

Faute transitoire.........................Faute temporaire externe.

Fiabilité.......................................Sûreté de fonctionnement selon le point de vue de la continuité du service délivré.

Mesure de la continuité de la délivrance d'un service correct ou, de façon équivalente, mesure du temps jusqu'à défaillance.

Fiabilitée stabilisée....................L'aptitude du système à délivrer un service correct est préservée (identité stochastique des temps jusqu'à défaillance successifs).

Fonction (d'un système).............Ce à quoi un système est destiné.

Fonction temps réel....................Fonction qui doit être remplie dans des intervalles de temps finis régis par l'environnement.

Intégrité......................................Fait de demeurer intact, donc d'éviter modifications ou suppressions indésirées.

Intrusion.....................................Faute externe opérationnelle intentionelle.

Logique maligne.........................Faute de conception intentionnelle.

Maintenabilité............................Facilité avec laquelle la maintenance d'un système peut être effectuée.

Mesure de la délivrance continue d'un service incorrect ou, de façon équivalente, mesure du temps jusqu'à restauration du service depuis la dernière défaillance survenue.

Maintenance corrective.............Actions entreprises durant la vie opérationnelle d'un système destinées à préserver ou améliorer son aptitude à délivrer un service en conformité avec la spécification.

Maintenance curative................Maintenance corrective destinée à éliminer des fautes ayant produit des erreurs qui ont été signalées.

Maintenance préventive............Maintenance corrective destinée à éliminer des fautes avant qu'elles ne produisent des erreurs.

Masquage de faute......................Effet résultant de l'application systématique de la compensation d'erreur, même en l'absence d'erreur.

Moyens (pour la sûreté de fonctionnement)....................Méthodes et techniques permettant de fournir à un système l'aptitude à délivrer un service conforme à la spécification, et de donner confiance dans cette aptitude.

Prévention des fautes, tolérance aux fautes, élimination des fautes, prévision des fautes.

Obtention (de la sûreté de fonctionnement)....................Méthodes et techniques destinées à fournir à un système l'aptitude à délivrer un service conforme au service conforme à la spécification.

Prévention des fautes et tolérance aux fautes.

Passivation (de faute)...............Actions destinées à empêcher une nouvelle activation d'une, ou de fautes.

Performabilité............................Mesure combinée performances-sûreté de fonctionnement.

Point de reprise.........................Instants durant l'exécution d'un processus dont l'état courant peut ultérieurement nécessiter d'être restauré.

Poursuite .... Forme de recouvrement d'erreur où la transformation de l'état erroné consiste à trouver un nouvel état à partir duquel le système peut encore fonctionner.

Prévention des fautes .... Méthodes et techniques destinées à empêcher l'occurrence ou l'introduction de fautes.

Prévision des fautes .... Méthodes et techniques destinées à estimer la présence, la création et les conséquences des fautes.

Recouvrement (d'erreur) .... Forme du traitement d'erreur où un état exempt d'erreur est substitué à l'état erroné.

Reprise .... Forme de recouvrement d'erreur où le système est ramené dans un état survenu avant l'occurrence d'erreur.

Restauration (du service) .... Transition de service incorrect à service correct.

Sécurité-innocuité .... Sûreté de fonctionnement selon le point de vue de la non-occurrence de défaillances catastrophiques.

Mesure du temps jusqu'à défaillance catastrophique.

Sécurité-confidentialité .... Sûreté de fonctionnement selon le point de vue de la prévention d'accès et/ou de manipulations non-autorisés de l'information.

Service .... Comportement d'un système, tel que perçu par son, ou ses utilisateurs.

Service correct .... Service délivré en conformité avec les spécifications du système.

Service incorrect .... Service délivré non en conformité avec la spécification.

Service temps réel .... Service qui doit être délivré dans des intervalles de temps finis régis par l'environnement.

Sévérité (d'une défaillance) .... Résultat de l'évaluation des conséquences du mode de défaillance sur l'environnement du système.

Spécification (d'un système) ...... Description agréée de la fonction ou du service attendu du système.

Structure (d'un système) .... Ce qui lui permet de faire ce qu'il fait.

Sûreté de fonctionnement .... Propriété qui permet aux utilisateurs d'un système de placer une confiance justifiée dans le service qu'il leur délivre.

Système .... Entité ayant interagi ou interféré, interagissant ou interférant, ou susceptible d'interagir ou d'interférer avec d'autres entités.

Ensemble de composants interconnectés en vue d'interagir.

Système à arrêt sur défaillance . Système dont toutes les défaillances sont, dans une mesure acceptable, des défaillances par arrêt.

Système à silence sur défaillance .... Système dont toutes les défaillances sont, dans une mesure acceptable, des défaillances par écrasement.

Système atomique .... Système dont la structure interne ne peut être discernée, ou n'est pas intéressante et peut être ignorée.

Système sûr en présence de défaillance .... Système dont toutes les défaillances sont, dans une mesure acceptable, des défaillances bénignes.

Système temps réel .... Système qui remplit au moins une fonction temps réel ou qui délivre au moins un service temps réel.

Test .... Vérification dynamique effectuée avec des entrées valuées.

Test aléatoire ou **statistique** .... Test où les jeux d'entrée sont sélectionnés selon une distribution probabiliste du domaine d'entrée.

Test basé sur des fautes .... Test destiné à révéler des classes de fautes spécifiques.

Test de conformité .... Test dont le but est de vérifier si le système satisfait ses spécifications.

Test déterministe .... Méthode de test où les jeux d'entrée sont déterminés par un choix sélectif selon le critère retenu.

Test fonctionnel .... Test où les jeux d'entrée sont sélectionnés selon des critères relatifs à la fonction du système.

Test de recherche de fautes .... Test destiné à révéler des fautes.

Test opérationnel........................Test destiné à évaluer la sûreté de fonctionnement, avec un profil d'entrée représentatif des conditions opérationnelles.

Test structurel...........................Test où les jeux d'entrée sont sélectionnés selon des critères relatifs à la structure du système.

Tolérance aux fautes.................Méthodes et techniques destinées à fournir un service conforme à la spécification en dépit des fautes.

Traitement d'erreur..................Opérations destinées à éliminer les erreurs, si possible avant qu'une défaillance ne survienne.

Traitement de faute...................Opérations destinées à éviter qu'une, ou des fautes ne soient activées à nouveau.

Utilisateur (d'un système).........Autre système (humain ou physique) qui interagit avec le système considéré.

Validation (de la sûreté de fonctionnement)....................Méthodes et techniques destinées à avoir confiance dans l'aptitude du système à délivrer un service conforme à la spécification.

Elimination des fautes et prévision des fautes.

Vérification..............................Processus consistant à déterminer si le système satisfait des propriétés, appelées conditions de vérification, qui peuvent être générales et indépendantes des spécifications, ou spécifiques et déduites des spécifications.

Vérification dynamique............Vérification comportant l'activation du système.

Vérification de non-régression...........................Vérification effectuée après correction, afin de s'assurer que l'élimination de faute n'a pas eu de conséquences indésirables.

Vérification statique.................Vérification effectuée sans activer le système.

INDEX FRANCAIS-ANGLAIS

Evitement des fautes .... Fault avoidance
Exécution symbolique .... Symbolic execution
Faute .... Fault
Faute accidentelle .... Accidental fault
Faute active .... Active fault
Fautes corrélées .... Related faults
Faute de conception .... Design fault
Faute dormante .... Dormant fault
Faute douce .... Soft fault
Faute dure, ou solide .... Hard, or solid, fault
Faute externe .... External fault
Faute humaine .... Human-made fault
Fautes indépendantes .... Independent faults
Faute intentionnelle .... Intentional fault
Faute intermittente .... Intermitent fault
Faute interne .... Internal fault
Faute opérationnelle .... Operational fault
Faute permanente .... Permanent fault
Faute physique .... Physical fault
Faute temporaire .... Temporary fault
Faute transitoire .... Transient fault
Fiabilité .... Reliability
Fiabilitée stabilisée .... Stable reliability
Fonction (d'un système) .... Function (system ~)
Fonction temps réel .... Real-time function
Intégrité .... Integrity
Intrusion .... Intrusion
Logique maligne .... Malicious logic
Maintenabilité .... Maintenability
Maintenance corrective .... Corrective maintenance
Maintenance curative .... Curative maintenance
Maintenance préventive .... Preventive maintenance
Masquage de faute .... Fault masking
Moyens (pour la sûreté de fonctionnement). Means (for dependability)
Obtention (de la sûreté de fonctionnement).. Procurement (of dependability)
Passivation (de faute) .... Passivation (fault ~)
Performabilité .... Performability
Point de reprise .... Recovery point
Poursuite .... Forward recovery

Prévention des fautes	Fault prevention
Prévision des fautes	Fault forecasting
Recouvrement (d'erreur)	Error recovery
Reprise	Backward recovery
Restauration (du service)	Restoration (service ~)
Sécurité-innocuité	Safety
Sécurité-confidentialité	Security
Service	Service
Service correct	Correct service
Service incorrect	Incorrect service
Service temps réel	Real-time service
Sévérité (d'une défaillance)	Severity (failure ~)
Spécification (d'un système)	Specification (system ~)
Structure (d'un système)	Structure (system ~)
Sûreté de fonctionnement	Dependability
Système	System
Système à arrêt sur défaillance	Fail-stop system
Système à silence sur défaillance	Fail-silent system
Système atomique	Atomic system
Système sûr en présence de défaillance	Fail-safe system
Système temps réel	Real-time system
Test	Testing
Test aléatoire ou statistique	Random, or statistical, testing
Test basé sur des fautes	Fault-based testing
Test de conformité	Conformance testing
Test déterministe	Deterministic testing
Test fonctionnel	Functional testing
Test de recherche de fautes	Fault-finding testing
Test opérationnel	Operational testing
Test structurel	Structural testing
Tolérance aux fautes	Fault tolerance
Traitement d'erreur	Error processing
Traitement de faute	Fault treatment
Utilisateur (d'un système)	User (system ~)
Validation (de la sûreté de fonctionnement)	Validation (dependability ~)
Vérification	Verification
Vérification dynamique	Dynamic verification
Vérification de non-régression	Regression verification
Vérification statique	Static verification

ZUVERLÄSSIGKEIT:

GRUNDKONZEPTE

UND TERMINOLOGIE

Vorbemerkung zur deutschen Übersetzung

Diese deutsche Übersetzung des englischen Originals wurde, soweit es möglich und sinnvoll war, in Anlehnung an die deutsche und die harmonisierte internationale Normung realisiert. Dabei wurden insbesondere folgende Richtlinien und Normen berücksichtigt:

- DIN V VDE 0801/01.90 Grundsätze für Rechner in Systemen mit Sicherheitsaufgaben
- DIN 40041/12.90 Zuverlässigkeit - Begriffe
- CEI/IEC 50 (191):1990 International Electrotechnical Vocabulary, Chapter 191: Dependability and quality of service (Internationales Elektrotechnisches Wörterbuch, Kapitel 191, Zuverlässigkeit und Dienstgüte).

Bei der deutschen Übersetzung ergaben sich dabei im wesentlichen die folgenden Problemkreise:

a) Die Übertragung der drei Begriffe 'fault', 'error' und 'failure' auf die im Deutschen hauptsächlich benutzten zwei Begriffe Fehler und Ausfall (bzw. Versagen) kann nur teilweise durch die Einführung von Fehlerursache für 'fault' und Fehlzustand für 'error' erleichtert werden. Insbesondere bei den zusammengesetzten Begriffen ergeben sich dann aber Schwierigkeiten, die auch mit der historischen Verwendung von Begriffen zusammenhängen. Soweit es kontextabhängig klar ist, ob über Fehlerursache oder Fehlzustand gesprochen wird, wird im folgenden daher Fehler verwendet.

b) Die Übersetzung der Begriffe 'safety' und 'security' erfolgte bisher gleichermaßen mit Sicherheit, wobei die unterschiedliche Ausprägung dieser Begriffe im Deutschen meist nur aus dem Kontext ersichtlich ist. Hier wird im folgenden der Begriff Sicherheit nur für die Übersetzung von 'safety' verwendet, wenn es also im wesentlichen um die Unfallvermeidung geht, während 'security' durch Vertraulichkeit übersetzt wird, auch wenn die 'security' mehr umfaßt als nur die Vertraulichkeit, z. B. Integrität, Verfügbarkeit, Zugriffsschutz.

c) Für die Übersetzung des zentralen Begriffs 'dependability' gab es die Möglichkeit, Verläßlichkeit oder Zuverlässigkeit zu wählen. Verläßlichkeit hat den Vorteil, daß die Erweiterung der Definition von Zuverlässigkeit um Sicherheit und Vertraulichkeit durch das neue Wort zum Ausdruck gebracht wird (Zuverlässigkeit [nach DIN 40041 und IEC 191] = Funktionsfähigkeit + Verfügbarkeit + Instandhaltbarkeit; Verläßlichkeit = Zuverlässigkeit + Sicherheit + Vertraulichkeit). Andererseits hat sich in der nationalen und internationalen Normung Zuverlässigkeit als Übersetzung von 'dependability' eingebürgert, wenn auch noch mit dem o. g. eingeschränkten Begriffsinhalt der Zuverlässigkeit. Damit ist aber im Deutschen bereits eine Begriffserweiterung der Zuverlässigkeit erfolgt, die früher nur die Übersetzung von 'reliability' war. Mit diesem Bericht kann vielleicht eine erneute Begriffserweiterung initiiert werden, die in Einklang mit der internationalen Entwicklung ist. Daher wurde hier die Übersetzung Zuverlässigkeit für 'dependability' gewählt.

Bei einigen Begriffen wird in diesem Bericht eine von der deutschen Normung abweichende Begriffserläuterung gegeben. Dies geschieht bewußt, um keine allzu große Diskrepanz zum englischen Original zu erzeugen.

Allgemein war aber das Bestreben bei der Übersetzung, keine rein wörtliche Übersetzung zu wählen, sondern die Ideen und Konzepte korrekt zu repräsentieren. Dabei ist zu berücksichtigen, daß die englische Version die Zustimmung eines größeren Arbeitskreises gefunden hat und daher ggf. als Referenz zu benutzen ist.

EINLEITUNG

Dieses Dokument hat zum Ziel, informelle aber dennoch präzise Definitionen für die verschiedenen Attribute zu geben, die die Zuverlässigkeit eines Rechensystems charakterisieren. Es ist ein Beitrag zu der Arbeit, die innerhalb der wissenschaftlichen und technischen Fachwelt "zuverlässiges und fehlertolerantes Rechnen" geleistet wird [Avi 67, Jes 7, Mel 77, Avi 78, Ran 78, Car 79, And 81, FTC 82, Sie 82, Cri 85a, Lab 85, Avi 86, Lap 89], um klare und auf breiter Basis akzeptierte Definitionen für einige grundlegende Konzepte vorzuschlagen.

Zunächst wird die Zuverlässigkeit als ein globales Konzept eingeführt, das die gewöhnlichen Attribute Funktionsfähigkeit, Verfügbarkeit, Sicherheit und Vertraulichkeit beinhaltet. Die grundlegenden Definitionen, die im ersten Abschnitt angegeben werden, werden in den folgenden Abschnitten kommentiert und um zusätzliche Definitionen erweitert. Im Anhang ist ein Glossar enthalten, das die Definitionen, die im Laufe des Textes angegeben werden, noch einmal zusammenfassend wiedergibt. Der Aufbau erfolgt so, daß keine Vorwärtsreferenz erforderlich ist. **Fettdruck** wird benutzt, wenn ein Begriff definiert wird, *Kursivdruck* bedeutet nur eine Hervorhebung, um die

Aufmerksamkeit des Lesers zu gewinnen. Die Richtlinien, die der gewählten Präsentation zugrundeliegen, können folgendermaßen zusammengefaßt werden:

- Suche nach einer geringen Zahl von Konzepten, die die Attribute der Zuverlässigkeit ausdrücken können.
- Benutzung von Begriffen, die - soweit wie möglich - identisch mit ihrer normalen Bedeutung verwendet werden oder so dicht wie möglich zu ihrer normalen Verwendung liegen; in der Regel hat ein Begriff, der nicht explizit definiert wird, seine normale Bedeutung behalten (wie in jedem Lexikon gegeben);
- Hervorhebung der Integration [Gol 82, Ran 86] (im Gegensatz zur Spezialisierung) durch die Unabhängigkeit der gegebenen Definitionen in bezug auf die Fehlerklassen.

Dieses Dokument kann als ein minimaler Konsens innerhalb der Fachwelt betrachtet werden, um fruchtbare Diskussionen zu ermöglichen; zusätzlich soll dieses Dokument dazu geeignet sein, a) von anderen Gremien benutzt zu werden (z. B. Normungsgremien), und b) für Ausbildungszwecke eingesetzt zu werden. Daher ist die zugehörige Bemühung um Begriffsdefinitionen noch nicht am Ende: Worte sind nur insoweit von Interesse, wie sie ein Konzept kennzeichnen und es ermöglichen, Ideen und Blickpunkte mitzuteilen und gemeinsam zu nutzen. Das Dokument hat nicht die Absicht, den gegenwärtigen Stand unveränderlich festzuschreiben: die Konzepte, die hier präsentiert werden, müssen sich mit der Technologie und mit unserem Fortschritt beim Verständnis und Beherrschen der Spezifikation, des Entwurfs und der Bewertung von zuverlässigen Rechensystemen weiterentwickeln.

Was in diesem Dokument dargestellt wird, würde ohne die vielen Diskussionen nicht existieren, die mit vielen Kollegen, insbesondere mit den Mitgliedern der IFIP WG 10.4, geführt wurden.

1. Grundlegende Definitionen

Die **Zuverlässigkeit**[1] wird definiert als die Vertrauenswürdigkeit eines Rechensystems, so daß Vertrauen in die Leistung, die es erbringt, gesetzt werden kann [Car 82]. Die **Leistung**, die von dem System erbracht wird, ist

---

[1] In DIN 40041/12.90 wird hingegen folgende Definition angegeben: Zuverlässigkeit ist die Beschaffenheit einer Einheit bezüglich ihrer Eignung, während oder nach vorgegebenen Zeitspannen bei vorgegebenen Anwendungsbedingungen die Zuverlässigkeitsforderungen zu erfüllen.

sein Verhalten, wie es von seinem Benutzer *beobachtet wird*; der **Benutzer** ist dabei ein anderes System (menschlich oder physikalisch), das mit dem vorhergehenden *zusammenwirkt.*

Abhängig von der Anwendung des Systems kann ein unterschiedliches Gewicht auf die verschiedenen Ausprägungen der Zuverlässigkeit gelegt werden, d. h. die Zuverlässigkeit kann aufgrund unterschiedlicher, aber komplementärer *Eigenschaften* betrachtet werden, und dementsprechend können die *Attribute* der Zuverlässigkeit definiert werden:

- in Hinblick auf das *Bereitsein zum Gebrauch* bedeutet zuverlässig **verfügbar**;
- in Hinblick auf die *Kontinuität der Leistung* bedeutet zuverlässig **funktionsfähig**;
- in Hinblick auf die *Vermeidung von katastrophalen Folgen für die Umgebung* bedeutet zuverlässig **sicher**;
- in Hinblick auf die *Verhinderung von nicht autorisiertem Zugriff* und/oder der *nicht autorisierten Handhabung von Informationen* bedeutet zuverlässig **geschützt**.

Ein **Ausfall** des Systems tritt auf, wenn die erbrachte Leistung nicht mehr mit der Spezifikation übereinstimmt. Dabei stellt die **Spezifikation** die akzeptierte Beschreibung der erwarteten Funktion und/oder Leistung des Systems dar. Ein **Fehler** ist der Teil des Systemzustands, der dafür verantwortlich ist, daß im folgenden ein Ausfall auftritt: ein Fehler, der die Leistung beeinträchtigt, ist ein Zeichen dafür, daß ein Ausfall auftritt oder aufgetreten ist. Die verantwortliche oder hypothetische Ursache für einen Fehler ('error') ist eine **Fehlerursache** ('fault').

Die Entwicklung von zuverlässigen Rechensystemen verlangt nach der gemeinsamen Verwendung einer Reihe von Methoden, die folgendermaßen klassifiziert werden können:

- **Fehlerverhinderung**: wie das Auftreten und die Einführung von Fehlern verhindert werden kann;
- **Fehlertoleranz**: wie eine Leistung zur Verfügung gestellt werden kann, die trotz Fehlern mit der Spezifikation übereinstimmt;
- **Fehlerbeseitigung**: wie die Anwesenheit von Fehlern (Anzahl und Schwere) reduziert werden kann;
- **Fehlervorhersage**: wie die augenblickliche Zahl, das zukünftige Auftreten und die Folgen von Fehlern abgeschätzt werden können.

Fehlerverhinderung und Fehlertoleranz können als Zuverlässigkeits**verfahren** betrachtet werden: wie kann ein System *mit der*

Möglichkeit ausgestattet werden, eine spezifikationsgemäße Leistung zu erbringen; Fehlerbeseitigung und Fehlervorhersage können als Zuverlässigkeits **validation** betrachtet werden: wie kann *Vertrauen* in die Fähigkeit des Systems *erreicht werden*, daß es eine spezifikationsgemäße Leistung erbringt.

Vertrauen in die Leistung eines Systems und die Rechtfertigung für dieses Vertrauen basieren auf der *Beurteilung* des Systems, die im wesentlichen in Hinblick auf die *Kenngrößen* der Zuverlässigkeit, die zur Leistung des Systems gehören, durchgeführt wird.

Die Begriffe, die bisher eingeführt wurden, lassen sich in drei Klassen einteilen (Abb. 1):

- die **Beeinträchtigungen** der Zuverlässigkeit: Fehlerursachen, Fehlzustände, Ausfälle; sie sind unerwünschte - aber in der Regel nicht unerwartete - Umstände, die Unzuverlässigkeit verursachen oder daraus resultieren (dabei ist die Definition von Unzuverlässigkeit einfach abgeleitet aus der Definition von Zuverlässigkeit: Vertrauen kann nicht, oder kann nicht mehr, in die Leistung gesetzt werden);
- die **Mittel** zur Zuverlässigkeit: Fehlerverhinderung, Fehlertoleranz, Fehlerbeseitigung, Fehlervorhersage; dies sind Methoden und Techniken, die es einem ermöglichen, a) die Fähigkeit bereitzustellen, eine Leistung zu erbringen, in die Vertrauen gesetzt werden kann, und b) Vertrauen in diese Fähigkeit selbst zu setzen.
- die **Kenngrößen** der Zuverlässigkeit: Verfügbarkeit, Funktionsfähigkeit, Sicherheit, Vertraulichkeit; a) ermöglichen diese, daß die Eigenschaften, die von dem System erwartet werden, ausgedrückt werden können, und b) gestatten sie, daß die Qualität des Systems, wie sie sich aufgrund der Beeinträchtigungen und der dagegen eingesetzten Mittel ergibt, beurteilt werden kann.

2. Einführung der Zuverlässigkeit als Grundkonzept

Eine natürliche Tendenz einer jeden wissenschaftlichen oder technischen Disziplin ist es, ihr Arbeitsgebiet in einem ersten Schritt zu beschränken, um dort - rasche - Fortschritte bei der Lösung der betreffenden Probleme zu erreichen. Bald kommt jedoch eine Zeit, in der das Zusammenwirken mit anderen Disziplinen nicht länger ignoriert werden kann. Man unterliegt dann

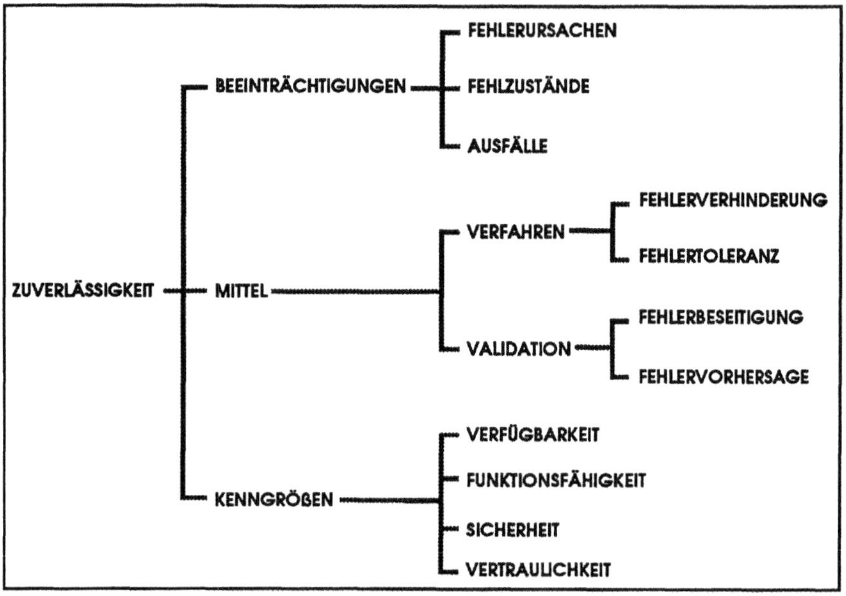

Abb. 1 - Der Zuverlässigkeitsbaum

leicht der Versuchung, diese anderen Disziplinen als "Spezialfälle" der eigenen Disziplin zu betrachten. Das hat normalerweise große Debatten zur Folge, die von den Anhängern der verschiedenen Disziplinen jeweils unter Benutzung der eigenen Fachsprache abgehalten werden. Dies war auch im Hinblick auf die Begriffe Funktionsfähigkeit, Sicherheit und Vertraulichkeit für den Bereich der Rechensysteme der Fall. Zunächst bestand die Hauptsorge darin, funktionierende Rechensysteme zu haben: *Funktionsfähigkeit* war gefordert. Als die Rechensysteme funktionsfähig wurden, wurde ihre Leistung regelmäßig benutzt und benötigt, womit *Verfügbarkeit* wichtig wurde. Die Verwendung von Rechensystemen in kritischen Anwendungsbereichen führte dazu, daß die *Sicherheit* an Bedeutung gewann. Das sicherste System ist oft ein System, das gar nichts tut, was aber wiederum nicht sehr hilfreich ist; daher betrachten Leute, für die Sicherheit im Vordergrund steht, die Funktionsfähigkeit als einen Unterbegriff der Sicherheit. Das Auftreten von verteilten Systemen verschlechterte die *Vertraulichkeit*. Ein geschütztes System wiederum soll Funktionen erfüllen. Zusätzlich können Verletzungen der Vertraulichkeit katastrophale Folgen haben. Daher betrachten Leute, für die Vertraulichkeit im Vordergrund steht, Sicherheit und Funktionsfähigkeit als Unterbegriffe der Vertraulichkeit.

Die Beziehungen zwischen Funktionsfähigkeit, Sicherheit und Vertraulichkeit sind allerdings komplizierter als einfache Abhängigkeiten.

Betrachten wir das Beispiel einer Fehlerursachen, die absichtlich in ein Rechensystem eingebaut wurde, um zu einem von einem "Terroristen" geplanten Zeitpunkt - und unter seiner Kontrolle - einen Systemausfall hervorzurufen, dessen Folgen von dem Benutzer zunächst nicht als kritisch angesehen werden, bis er die Ausfallursachen erkennt. Dieses Beispiel beinhaltet auf jeden Fall Funktionsfähigkeit, Sicherheit und Vertraulichkeit in einer verwickelten und wechselnden Art, abhängig von dem jeweiligen Standpunkt[2]. Auf jeden Fall gilt, daß der Benutzer in die Leistung, die ein solches System erbringt, das nicht zuverlässig im ursprünglichen Sinne des Wortes ist, kein Vertrauen setzen kann oder sollte.

Die vorangegangene Diskussion zeigt deutlich, daß es nicht die Absicht dieses Dokuments ist, zu der Kontroverse beizutragen, ob Funktionsfähigkeit der Sicherheit übergeordnet ist oder umgekehrt, und ebenso bezogen auf Vertraulichkeit. Entscheidend ist in der Beziehung zwischen Funktionsfähigkeit, Sicherheit und Vertraulichkeit einerseits und dem Begriff Zuverlässigkeit andererseits, daß die Erstgenannten Kenngrößen des Zuletztgenannten sind: und in dieser Beziehung ist es wünschenswert, daß eine gegenseitige Befruchtung stattfinden kann. Dies ist der Hauptgrund, warum ein weiteres Wort zu der bereits langen Liste von Begriffen - Funktionsfähigkeit, Verfügbarkeit, Sicherheit, Vertraulichkeit, usw. - hinzugefügt wird.

3. SYSTEMFUNKTION, -VERHALTEN, -STRUKTUR UND -SPEZIFIKATION

Bis jetzt haben wir ein **System** - implizit - als ein Ganzes betrachtet und sein extern beobachtbares Verhalten betont. Eine mit dieser Black-Box-Betrachtung übereinstimmende Definition ist die folgende: eine Einheit, die mit anderen Einheiten zusammengewirkt hat, zusammenwirkt oder wahrscheinlich zusammenwirken wird. Diese anderen Systeme waren, sind oder werden Bestandteil der **Umgebung** des betrachteten Systems sein[3]. Ein

---

[2] Es ist ebenso bemerkenswert, daß die Ereignisse, die innerhalb der Rubrik "Risiken von Rechensystemen für die Öffentlichkeit" in den ACM Software Engineering Notes veröffentlicht werden, sich auf Funktionsfähigkeit, auf Sicherheit und auf Vertraulichkeit beziehen.

[3] a) Rekursive Definitionen werden nicht der Rekursion wegen angegeben. Ziel ist vielmehr, die Abhängigkeit in bezug auf den Blickwinkel zu betonen. Das gilt insbesondere für die Abgrenzung des Begriffs System: die Grenzen eines gegebenen Systems können in Abhängigkeit von dem Betrachter variieren: ist es der Entwickler, der Benutzer, die Wartungsmannschaft, usw.

b) Die Vergangenheits-, Gegenwarts- und Zukunftsform wird benutzt, um darzulegen, daß die Umgebung eines Systems sich mit der Zeit ändern kann, insbesondere in Hinblick

System-**Benutzer** ist Teil der Umgebung, die mit dem betrachteten System *zusammenwirkt*: der Benutzer versorgt das System mit Eingaben und/oder erhält Ausgaben vom System, und sein Kennzeichen ist, daß er die *Leistung benutzt*, die das System bereitstellt.

Die **Funktion** eines Systems ist das, wofür das System *gedacht* ist [Kui 85]. Das **Verhalten** eines Systems ist das, was ein System *macht*. Die Struktur des Systems ist das, was *dazu führt, daß ein System das macht, was es macht* [Zie 76]. Wenn man den Gedanken von [And 81] folgt, dann ist ein System, von einem strukturellen Blickpunkt aus, eine Menge von Komponenten, die verbunden sind, um zusammenzuwirken; eine **Komponente** ist ein weiteres System, usw. Die Rekursion endet, wenn ein System als **atomar** betrachtet wird: jede weitere interne Struktur kann nicht wahrgenommen werden, oder sie ist nicht von Interesse und kann daher ignoriert werden. Der Begriff 'Komponente' muß in einem breiten Sinne verstanden werden: Schichten eines Systems genauso wie Komponenten zwischen den Schichten; da eine Komponente auch wieder ein System ist, enthält es auch die Abhängigkeiten zwischen den Komponenten, aus denen es besteht. Eine eher klassische Definition der System-Struktur ist, was ein System *ist*. Solch eine Definition paßt perfekt, wenn ein System ohne Rücksicht auf die Mängel seiner Zuverlässigkeit betrachtet wird, d. h. in Fällen, wo die Struktur als *fest* angesehen wird. Wir wollen uns allerdings nicht auf Systeme einschränken, deren Struktur fest ist. Insbesondere müssen wir strukturelle Veränderungen zulassen, die von Zuverlässigkeitsmängeln verursacht wurden oder daraus resultieren. Daraus folgt, daß eine Struktur Zustände haben kann[4]. Daraus ergibt sich eine Definition für den Begriff **Zustand**: Beschaffenheit einer Einheit mit Hinblick auf eine Menge von Randbedingungen, *sowohl bezogen auf das Verhalten als auch auf die Struktur*[5].

Von seiner ursprünglichen Definition her (dem vom Benutzer beobachteten Verhalten) ist die Leistung, die ein System erbringt, eindeutig eine *Abstraktion*

---

auf die verschiedenen Phasen des Lebenszyklus vom System. Z. B. paßt der Begriff Programmentwicklungsumgebung in diese Definition ebenso wie die physikalische Umgebung des Systems während des Betriebs.

4 a) Man kann daher sagen, daß eine 'Struktur' auch ein 'Verhalten' hat, insbesondere bezogen auf die Zuverlässigkeitsmängel, selbst wenn die betrachteten Geschwindigkeiten der Evolution mit Hinblick auf i) die Anforderungen des Benutzers auf der einer Seite, und ii) die Mängel auf der anderen Seite - hoffentlich - unterschiedlich sind.

b) Die angegebene Definition erlaubt es, andere Arten von Systemen mit unterschiedlichen Strukturen einzuschließen, z. B. adaptive - insbesondere wissensbasierte - Systeme.

5 Diese Definition zielt darauf ab, die Relativität des Begriffs Zustand hervorzuheben, die direkt von den betrachteten Phänomenen und Randbedingungen abhängt; z. B. Zustand mit Bezug auf die Rechenaktivitäten, Zustand mit Bezug auf Auftreten von Ausfällen.

des Systemverhaltens. Es ist bemerkenswert, daß diese Abstraktion sehr stark von der Anwendung, die das Rechensystem unterstützt, abhängt. Ein Beispiel für diese Abhängigkeit ist die entscheidende Rolle, die die Zeit in dieser Abstraktion spielt: Die Zeit-Granularität des Systems und die des Benutzers (der Benutzer) sind im allgemeinen unterschiedlich, und der Unterschied variiert von einer Anwendung zur anderen. Außerdem ist die Leistung natürlich nicht auf die Ausgaben eingeschränkt, sondern beinhaltet alle Wechselwirkungen, die für den Benutzer von Interesse sind; z. B. ist die Abfrage von Sensoren eindeutig ein Teil der Leistung, die von einem Überwachungssystem erwartet wird.

Bis jetzt haben wir nur die Einzahl für Funktion und Leistung benutzt. Normalerweise erfüllt ein System aber mehr als eine Funktion, und es stellt mehr als eine Leistung zur Verfügung. Die Funktion bzw. die Leistung kann als eine Zusammensetzung von Einzelfunktionen bzw. Einzelleistungen betrachtet werden. Aus Gründen der Vereinfachung werden wir die Pluralform - Funktionen und Leistungen - benutzen, wenn es wichtig ist, zwischen verschiedenen Funktionen und Leistungen zu unterscheiden.

Von besonderem Interesse in Hinblick auf die Zuverlässigkeit sind die zeitbezogenen Eigenschaften. Eine **Echtzeit-Funktion oder -Leistung** ist eine Funktion oder Leistung, von der gefordert ist, daß sie innerhalb endlicher Zeitintervalle, die *von der Umgebung diktiert* werden, erbracht wird, und ein **Echtzeit-System** ist ein System, das zumindest eine Echtzeit-Funktion erfüllt oder zumindest eine Echtzeit-Leistung erbringt [PDC 90].

Die *Spezifikation* eines Systems beschreibt die Erwartungen an ein System in Form von a) der erwarteten Funktion und/oder der erwarteten Leistung, und b) Bedingungen, unter denen sie erfüllt werden müssen: Umgebung, Zeitintervall, Leistungserbringung, Beobachtbarkeit, usw. Die Funktion und/oder die Leistung werden normalerweise zunächst in Hinblick darauf spezifiziert, was in bezug auf die Hauptzielrichtung des Systems erfüllt oder erbracht werden *sollte*. Wenn sicherheitsbezogene oder vertraulichkeitsbezogene Systeme betrachtet werden, wird diese Spezifikation normalerweise um den Aspekt, was *nicht* passieren *sollte*, ergänzt (z. B. die gefährlichen Zuständen, von denen eine Katastrophe droht, oder das Zugänglichmachen von sensiblen Informationen). Eine derartige Spezifikation kann wiederum zur Spezifikation von - weiteren - Funktionen und Leistungen führen, die das System erfüllen oder erbringen sollte, um die Wahrscheinlichkeit für das, was nicht eintreten *sollte*, zu verringern (z. B. Überprüfung der Zugriffsrechte von einem Benutzer und Identifizierung desselben).

Weiterhin können diese verschiedenen Spezifikationen:

a) in unterschiedlichem Detaillierungsgrad verfaßt sein: Anforderungsspezifikation, Entwurfsspezifikation, Implementierungsspezifikation, usw.

b) entsprechend der Abwesenheit oder Anwesenheit von Komponentenausfällen verfeinert sein; der erste Fall bezieht sich auf den meist *Normal*betriebsfall genannten Fall, während der zweite Fall sich auf den Betriebsfall *mit reduzierter Leistung* beziehen kann, falls die nicht ausgefallenen Betriebsmittel nicht mehr ausreichen, um die normale Leistung zu erbringen.

Daraus folgt, daß es normalerweise nicht eine einzige Spezifikation gibt, sondern mehrere, und daß deshalb ein System in bezug auf einige dieser Spezifikationen ausfallen kann, andere aber noch immer erfüllt.

Es ist entscheidend, daß sich zwei Personen oder Parteien auf eine Spezifikation geeinigt haben: der Systemersteller (im weitesten Sinne des Begriffs: Entwerfer, Erbauer, Verkäufer, usw.) und der (menschliche) Systembenutzer[6]. Die Einigung ist notwendig, damit die Spezifikation als Basis für die Beurteilung dienen kann, ob eine erbrachte Leistung akzeptabel ist oder nicht, oder gleichbedeutend damit, ob ein Ausfall stattgefunden hat oder nicht. Was bei einem vorgegebenen Detaillierungsgrad bezogen auf eine Spezifikation als eine akzeptable Leistung bezeichnet werden kann, kann nicht mit der Spezifikation auf einem geringeren Detaillierungsgrad übereinstimmen, falls Fehler bei der Verfeinerung der Spezifikation aufgetreten sind, die zu Spezifikationsfehlern geführt haben; Spezifikationsfehler wiederum können irgendeine andere Spezifikation betreffen. Allgemeiner gesagt, eine Spezifikation kann nicht als unveränderlich betrachtet werden, sobald sie erstellt worden ist. Dies wäre sonst eine Verdrängung der Lebenserfahrung, daß Änderungen unabwendbar sind. Die Änderungen können in Modifikationen der System-Anforderungen begründet sein: Modifikationen der erwarteten Funktion und/oder Leistung, oder Korrektur einiger Fehler[7]. Wichtig ist, daß man sich auf die Spezifikation (wieder) geeinigt hat.

---

6 Die Einigung kann implizit sein, z. B. wenn ein System gekauft wird, das mit seiner Spezifikation und den Benutzer-Handbüchern ausgeliefert wird, oder bei der Benutzung eines Standardsystems (Massenprodukts).

7 Wir sind daher mit einem Kreisproblem konfrontiert: eine Referenz wird für die Beurteilung, ob die erbrachte Leistung akzeptabel ist oder nicht, benötigt, und diese Referenz kann fehlerhaft sein. Der Verbesserung der Spezifikationen wurde lange Zeit viel Aufmerksamkeit geschenkt, darunter auch Vorschläge für Lebenszyklus-Modelle, die dies zum Ziele hatten [Boe 88].

Auf der Grundlage des o. g. Blickpunkts der System-Struktur treffen der Begriff der Funktion, der Leistung und der Spezifikation in natürlicher Weise genauso auf die Komponenten zu. Dies ist besonders im Entwurfsprozeß von Interesse, wenn Standardkomponenten, Software oder Hardware, benutzt werden: die Funktion und/oder die Leistung, die sie erbringen können, ist wichtiger für den Entwickler als ihr detailliertes (internes) Verhalten.

4- Beeinträchtigungen der Zuverlässigkeit

4.1- Fehlerursachen[8]

Fehler und ihre Quellen sind extrem unterschiedlich. Sie können nach drei Hauptkriterien klassifiziert werden, nach ihrer Art, ihrem Ursprung und ihrer Dauer.

Die *Art* der Fehler führt zur folgenden Unterscheidung:

- **Zufallsfehler** (unabsichtlicher Fehler, zufällige Fehlerursache), die zufällig auftreten oder erzeugt werden;
- **Absichtsfehler** (absichtliche Fehlerursache), die absichtlich erzeugt werden, wahrscheinlich in böswilliger Absicht.

Der *Ursprung* von Fehlern kann wiederum nach drei Gesichtspunkten aufgeteilt werden:

- nach *phänomenologischen Gründen*, die zur folgenden Unterscheidung führen [Avi 78]:
  - **physikalische Fehler**, die auf physikalischen Phänomenen beruhen,
  - **menschliche Fehler**, die auf der menschlichen Unzulänglichkeit basieren;
- nach den *Systemgrenzen*, die zur folgenden Unterteilung führen:
  - **interne Fehler**, die solche Teile des Systemzustandes sind, die bei Aufruf durch eine Rechenaktivität einen Fehler hervorrufen,
  - **externe Fehler**, die von der Beeinflussung des Systems durch seine physikalische Umgebung (elektromagnetische Störungen, Strahlung, Temperatur, Erschütterung, usw.) oder vom Zusammenwirken mit seiner menschlichen Umgebung hervorgerufen werden;

---

8 In diesem Kapitel steht Fehler in der Regel, insbesondere bei zusammengesetzten Begriffen, für Fehlerursache, englisch 'fault'.

- nach der *Phase der Entstehung* in bezug auf den Lebenszyklus des Systems, was zu folgender Einteilung führt:
  - **Entwurfsfehler**, die auf Mängeln beruhen, entweder a) während der Entwicklung des Systems (von der Anforderungsspezifikation bis zur Implementierung) oder während der folgenden Modifikationen, oder b) während des Erstellens von Prozeduren für das Betreiben oder die Wartung des Systems;
  - **Benutzungsfehler (Betriebsfehler)**, die während der Nutzung des Systems auftreten.

Eine Unterscheidung kann ebenso in Hinblick auf die *Dauer* der Fehler gemacht werden, was zu der Einteilung führt:

- **Permanente Fehler**, deren *Anwesenheit* nicht von einer punktuellen Bedingung abhängig ist, sei sie intern (Rechenaktivität) oder extern (Umgebung),
- **Temporäre Fehler**, deren Anwesenheit von einer punktuellen Bedingung abhängt und die daher nur während einer bestimmten Zeit vorhanden sind.

Vertraulichkeitsthemen sind dominiert von absichtlichen Fehlern (aber nicht eingeschränkt auf sie), die eindeutig *menschliche* Fehler sind. Absichtliche Fehler können entweder interne oder externe Fehler sein; typische Beispiele sind:

- für interne Fehler der Einbau von **bösartig fehlerhafter Logik** (z. B. sogenannte "Trojanische Pferde"), was einem absichtlichen *Entwurfsfehler* entspricht;
- für externe Fehler das **Eindringen** oder den illegalen Systemzugang als ein absichtlicher *externer* Fehler während des *Betriebs*.

Um erfolgreich zu sein, können absichtliche Fehler Vorteil aus zufälligen Fehlern ziehen, z. B. kann ein Eindringen einen Sicherungsmangel ausnutzen, der auf einem zufälligen Entwurfsfehler beruht; es gibt interessante und offensichtliche Ähnlichkeiten zwischen dieser Möglichkeit und einem zufälligen externen temporären Fehler, der den Mangel an Sicherungsvorkehrung ausnutzt.

Es kann argumentiert werden, daß die Einführung von phänomenologischen Gründen in den Klassifikationskriterien von Fehlern zu einem rekursiven endlosen Weg zurück führen kann, z. B. "Warum machen Programmierer Fehler?", "Warum versagen integrierte Schaltungen?". Der engere Begriff der Fehlerursache ist willkürlich und bietet eine Möglichkeit, die Rekursion zu beenden. Daher die gegebene Definition: *anerkannte oder angenommene*

Ursache für einen Fehlzustand. Diese Ursache kann sich mit dem gewählten Blickwinkel ändern: Fehlertoleranzmechanismen, Instandhaltungsingenieure, Reparaturwerkstatt, Entwickler, Halbleiter-Physiker, usw. Unserer Meinung nach endet die Rekursion *bei der Ursache, die verhindert oder toleriert werden sollte.* Dieser Gesichtspunkt sorgt für Konsistenz mit der Unterscheidung zwischen menschlichen und physikalischen Fehlerursachen: ein Rechensystem ist ein menschliches Machwerk, und somit ist jeder Fehler, der darin enthalten ist oder der es betrifft, letztendlich menschlichen Ursprungs, da er aus der menschlichen Unzulänglichkeit resultiert, alle Phänomene zu beherrschen, die das Verhalten des Systems betreffen. Im strengen Sinne kann eine Unterscheidung zwischen physikalischen und menschlichen Fehlerursachen (insbesondere Entwurfsfehler) als unnötig gehalten werden; dennoch ist diese Unterscheidung wichtig, wenn man die (augenblicklichen) Methoden und Techniken zur Erreichung und Validation von Zuverlässigkeit betrachtet. Wenn die oben erwähnte Rekursion nicht beendet wird, *ist eine Fehlerursache nichts anderes als die Folge des Ausfalls eines anderen Systems* (einschließlich des Entwicklers), *das für das betrachtete System eine Leistung erbracht hat oder jetzt erbringt.*

Beispiele für die vorangegangene Diskussion folgen:

- ein Entwurfsfehler resultiert von einem Versagen des Entwerfers;
- ein physikalischer interner Fehler wird durch einen Ausfall einer Hardware-Komponente verursacht. Dieser Fehler ist die Folge eines Fehlers auf der elektrischen oder elektronischen Ebene (die Fachwelt der "Zuverlässigkeit der Physik" bezeichnet Ausfälle selten als "plötzlich und unvorhersehbar"). Physikalisch-chemische Störungen können zurückverfolgt werden auf Fehler in der Hardware-Herstellung oder auf unsere mangelnde Kenntnisse der Halbleiterphysik; physikalische interne Fehlerursachen können als *wiederauftretende* Fehlerursachen betrachtet werden, solange ihre Reparatur aus dem Austauschen der ausgefallenen Komponente durch eine identische - noch nicht ausgefallene - Komponente besteht: die neue Komponente kann in der Zukunft genauso ausfallen, es sei denn, die Ausfallursache ist in den Entwurf und die Produktion zurückverfolgt und beseitigt worden, was zu einer modifizierten Komponente mit einer reduzierten Ausfallwahrscheinlichkeit führt;
- ein physikalischer oder menschlicher externer Fehler ist in der Tat ein Entwurfsfehler: die Unfähigkeit, alle Situationen vorherzusehen, die das System während des Betriebs durchlaufen muß, oder die Weigerung, einige von ihnen zu beachten (z. B. aus wirtschaftlichen Erwägungen); z. B.

- im Fall der elektromagnetischen Störungen: ist es ein externer Fehler oder ein Entwurfsfehler, z. B. der Mangel an ausreichender Abschirmung?
- im Fall eines Ausfalls, der durch einen Bediener hervorgerufen wird, der einen einzigen Buchstaben falsch tippt: ist es ein Bedienungsfehler oder ein Entwurfsfehler, wie der Mangel, daß das System nicht nach einer Bestätigung der Eingabe fragt [Nor 83]?

Der Blickwinkel der zeitlichen Dauer verdient die folgenden Kommentare:

1) Temporäre externe Fehler, die von der physikalischen Umgebung stammen, werden oft **transiente Fehler** genannt.
2) Temporäre interne Fehler werden oft **intermittierende Fehler** genannt; solche Fehler resultieren aus der Anwesenheit von selten auftretenden Kombinationen von Bedingungen; Beispiele sind a) inhaltsabhängige Fehler in Halbleiter-Speichern, Veränderungen in den Parametern einer Hardware-Komponente (Effekte der Temperatur-Veränderungen, Verzögerungen im Zeitverhalten aufgrund von parasitärer Kapazität, usw.) oder b) Situationen - sowohl in der Hardware als auch in der Software - , die auftreten, wenn die Systembelastung ein bestimmtes Maß übersteigt, wie z. B. marginales Zeitverhalten und Synchronisation. Der Begriff intermittierender Fehler ist im absoluten Sinne willkürlich: solche Fehler sind nichts anderes als (permanente) Fehler, deren Aktivierungsbedingung nicht einfach zu reproduzieren ist oder die selten genug auftritt; dennoch ist ihre Betrachtung sinnvoll, wie bereits bei der Unterscheidung zwischen physikalischen Fehlern und Entwurfsfehlern erwähnt wurde.

Aus dieser Diskussion ersieht man, daß *jeder Fehler als ein permanenter Entwurfsfehler betrachtet werden kann.* Dies ist in der Tat im absoluten Sinne wahr, aber es ist für den Entwickler oder den Betreiber eines Systems nicht sehr hilfreich.

Abbildung 2 faßt die verschiedenen Klassen von Fehlerursachen zusammen, die bisher behandelt wurden, unter Berücksichtigung der verschiedenen erwähnten Blickwinkel.

Wenn alle Kombinationen von Fehlerklassen entsprechend den fünf Blickwinkeln in Abbildung 2 möglich wären, gäbe es 32 verschiedene Fehlerklassen.

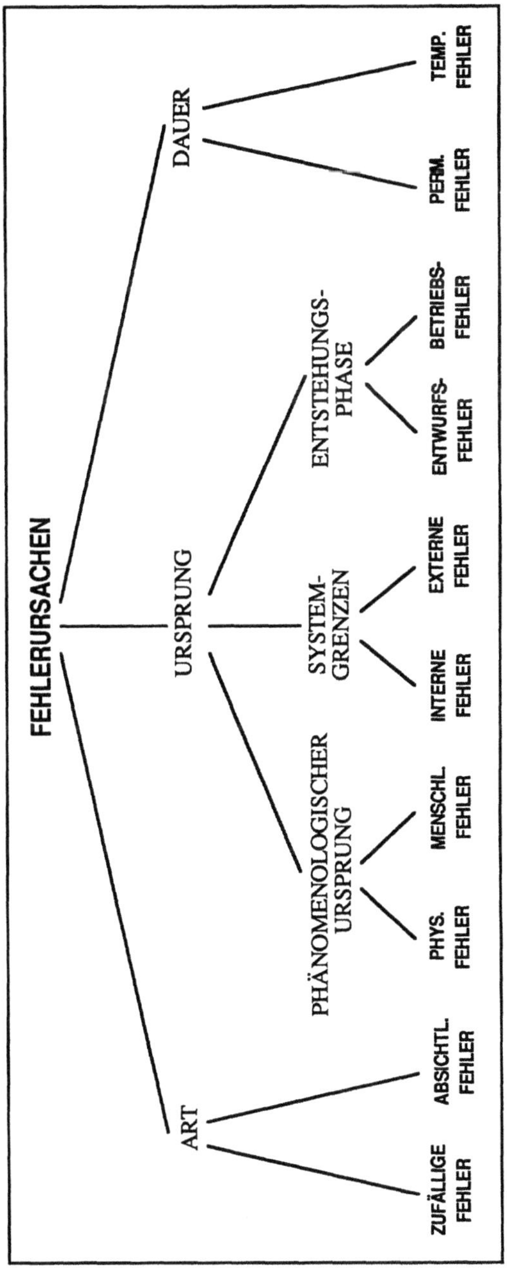

Abb. 2 - Die Fehlerklassen aus unterschiedlichen Blickwinkeln

In Wirklichkeit ist die Zahl der wahrscheinlichen Kombinationen geringer: Elf Kombinationen sind in den Zeilen von Abbildung 3 angegeben, die auch die normalerweise benutzten Namen dieser Kombinationen enthalten - *nicht ihre Definitionen*. Diese Namen werden allgemein benutzt, um in einer kompakten Form das Ergebnis der Kombination von verschiedenen Blickwinkeln auszudrücken.

4.2 - Fehlzustand

Die Definition eines Fehlzustands besagt, daß er *verantwortlich* ist für einen folgenden Ausfall. Ob ein Fehlzustand wirklich zu einem Ausfall führt oder nicht, hängt im wesentlichen von drei Gründen ab:

1) Vom Systemaufbau und insbesondere der Natur der existierenden Redundanz:
   - *beabsichtigte* Redundanz (eingeführt, um Fehlertoleranz bereitzustellen), die explizit dazu da ist, daß ein Fehlzustand nicht zu einem Ausfall führt,
   - *unbeabsichtigte* Redundanz (es ist praktisch sehr schwierig, wenn nicht sogar unmöglich, ein System ohne irgendeine Form der Redundanz zu bauen[9]), die die gleiche - aber unerwartete - Wirkung wie beabsichtigte Redundanz haben kann.
2) Von der Systemaktivität: ein Fehlzustand kann überschrieben sein, bevor er Unheil anrichten kann.
3) Von der Definition eines Ausfalls aus der Sicht des Benutzers: was für den einen Benutzer ein Ausfall sein kann, ist für einen anderen Benutzer höchstens eine Unannehmlichkeit. Beispiele sind a) die Zeit-Granularität des Benutzers: ein Fehlzustand, der durch die Schnittstelle zwischen System und Benutzer durchgelassen wird, kann oder kann nicht als ein Ausfall angesehen werden, je nach der Zeit-Granularität des Benutzers; b) die Notation der 'akzeptablen Fehlerrate' - implizit bevor ein Fehlzustand aufgetreten ist - in der Datenübertragung. Diese Diskussion erklärt, warum es oft wünschenswert ist, in der Spezifikation solche Bedingungen wie maximale Ausfallzeit (bezogen auf die Zeitgranularität des Benutzers) explizit zu erwähnen.

---

[9] Ein klassisches Problem beim Hardware-Testen ist die Beseitigung von 'falschen Redundanzen', deren Effekt die Maskierung von Fehlerursachen sein kann, wodurch sie die Erzeugung von Testmustern erschweren.

Art		Ursprung						Dauer		Normale Benennung
		Phänomenologische Ursache		Systemgrenzen		Entstehungsphasen				
Zufällige Fehler	Absichtl. Fehler	Phys. Fehler	Menschl. Fehler	Interne Fehler	Externe Fehler	Entwurfs-fehler	Benutzungs-fehler	Perm. Fehler	Temp. Fehler	
✓		✓		✓			✓	✓		Physikalische Fehler
✓		✓			✓		✓	✓		
✓		✓			✓		✓		✓	Transiente Fehler
✓		✓		✓			✓		✓	Intermittierende Fehler
✓			✓	✓		✓			✓	
✓			✓	✓		✓		✓		Entwurfsfehler
✓			✓		✓		✓		✓	Bedienungs fehler
	✓		✓	✓		✓		✓		Bösartig fehlerhafte Logik
	✓		✓	✓		✓			✓	
	✓		✓		✓		✓	✓		Illegaler Systemzugang
	✓		✓		✓		✓		✓	

Abb. 3 - Die Fehlerklassen entsprechend den Kombinationen von verschiedenen Blickwinkeln aus Abb. 2

4.3 - Ausfälle

Aufgrund der gegebenen Definition einer Systemstruktur ist die Diskussion, ob ein 'Ausfall' sich auf ein System oder eine Komponente bezieht, irrelevant, da eine Komponente auch wiederum ein System ist. Wenn wir es mit atomaren Systemen zu tun haben, ist der Begriff eines 'elementaren' Ausfalls angebracht.

Ein System kann nicht - und wird auch in der Regel nicht - immer in gleicher Weise ausfallen. Die Art und Weise, wie ein System ausfallen kann, sind die Ausfall-*Arten*, die nach drei Gesichtspunkten charakterisiert werden können: Bereich, Beobachtung durch den Benutzer und Auswirkungen auf die Umgebung.

Der Gesichtspunkt des *Ausfallbereichs* führt zu folgender Unterscheidung:

- **Werte-Ausfälle**: die Werte der bereitgestellten Leistung stimmen nicht mit der Spezifikation überein.
- **Zeit-Ausfälle**: das Zeitverhalten der erbrachten Leistung stimmt nicht mit der Spezifikation überein.

Solche allgemeinen Definitionen (keine Übereinstimmung mit der Spezifikation) gelten für **willkürliche Ausfälle**. Spezifischere Ausfallarten können unterschieden werden. Zum Beispiel kann der Begriff der Zeit-Ausfälle verfeinert werden in *verfrühte* Zeit-Ausfälle und *verspätete* Zeit-Ausfälle in Abhängigkeit davon, ob die Leistung zu früh oder zu spät erbracht wurde. Eine Klasse von Ausfällen, die sich sowohl auf die Werte-Ausfälle als auch auf die Zeit-Ausfälle bezieht, sind die **Stillstand-Ausfälle**: eine System-Aktivität, sofern sie noch existiert, ist für den Benutzer nicht mehr beobachtbar, und ein konstanter Wert wird als Leistung erbracht; der konstante Wert kann je nach Anwendung variieren, z. B. der letzte korrekte Wert, ein vorher definierter Wert, usw. Ein Spezialfall von Stillstand-Ausfällen sind die **Auslassungsausfälle** [Cri 85b, Ezh 86]: keine Leistung wird erbracht. Ein derartiger Ausfall kann als ein allgemeiner Grenzfall sowohl für den Werte-Ausfall (kein Wert) als auch für den Zeit-Ausfall (unendlich verspätet) angesehen werden; ein *permanenter* Auslassungsausfall ist ein **Total-Ausfall**. Ein System, dessen Ausfälle in der Regel nur Auslassungsausfälle sind, ist ein **Fail-stop-System**, und ein System, dessen Ausfälle in der Regel nur Total-Ausfälle sind, ist ein **Fail-silent-System** [Pow 88][10].

---

10 Der Begriff eines fail-stop-*Prozessors*, wie er in [Such 83] definiert ist, kann im Kontext von verteilten Systemen als ein Beispiel für ein fail-silent-System gesehen werden.

Wenn ein System mehrere Benutzer hat, führt der Gesichtspunkt der *Ausfall-Beobachtung* zu folgender Unterscheidung:

- **konsistente Ausfälle**: alle Systembenutzer beobachten die Ausfälle in gleicher Weise;
- **inkonsistente Ausfälle**: ein Ausfall kann für die verschiedenen Benutzer des Systems unterschiedlich beobachtbar sein; inkonsistente Ausfälle werden normalerweise nach [Lam 82] *byzantinische Ausfälle* genannt.

Der Ausfall-**Schweregrad** resultiert aus einer Bewertung der *Auswirkungen des Ausfalls* auf die Umgebung des Systems. Er ermöglicht es, die Ausfall-Arten zu ordnen. Ein interessanter Spezialfall sind solche Systeme, deren Ausfälle in zwei Gruppen mit erheblich unterschiedlichen Auswirkungen eingeteilt werden können:

- **unkritische Ausfälle**, bei denen die Auswirkungen der Ausfälle in derselben Größenordnung liegen (allgemein auf der Basis von Kosten) wie der Nutzen, der durch die Leistungserbringung ohne Ausfälle erreicht wird;
- **kritische Ausfälle**, bei denen die Auswirkungen erheblich größer sind als der Nutzen, der durch die Leistungserbringung ohne Ausfälle erreicht wird.

Ein System, dessen Ausfälle nur - oder zumindest zu einem akzeptablen Grad - unkritische Ausfälle sind, ist ein **Fail-safe System**. Der Begriff der Schwere eines Ausfalls ermöglicht es, den Begriff der Kritikalität zu definieren: die **Kritikalität** eines Systems bezeichnet die schlimmstmögliche Auswirkung eines (möglichen) Ausfalls[11]. Die Beziehung zwischen Ausfallarten und Ausfallschwere ist stark anwendungsabhängig. Dennoch existiert eine breite Klasse von Anwendungen, in denen Passivität als ein sicherer Betriebszustand betrachtet wird (z. B. bei Eisenbahn, Energieproduktion). Deshalb wird oft eine direkte Gleichsetzung zwischen fail-stop und fail-safe gemacht [Min 67, Nic 89].

---

[11] Zum Beispiel: das Kritikalitätsniveau, das von der Luftfahrt akzeptiert wird, ist folgendermaßen definiert [RTC 85]:

- kritisch: Funktionen, deren Ausfall die Fortführung eines sicheren Fluges und die Landung des Flugzeugs verhindern würde;
- wichtig: Funktionen, deren Ausfall die Fähigkeiten des Flugzeugs oder die Möglichkeit der Mannschaft, mit ungünstigen Flug-Bedingungen fertigzuwerden, reduziert;
- nicht-wichtig: Funktionen, deren Ausfall keine signifikante Verringerung der Fähigkeiten des Flugzeugs oder der Möglichkeiten der Mannschaft bedeutet.

4.4- Fehlerpathologie

Die Erzeugungs- und Offenbarungsmechanismen von Fehlerursachen, Fehlzuständen und Ausfällen können folgendermaßen zusammengefaßt werden:

1) Eine Fehlerursache ist **aktiv**, wenn sie einen Fehlzustand erzeugt. Eine aktive Fehlerursache ist entweder a) eine interne Fehlerursache, die vorher **inaktiv** war und die durch den Rechenprozeß aktiviert wurde (unter Einschluß des gleichzeitigen Existierens von Fehlerursachenbedingungen für eine intermittierende Fehlerursache), oder b) eine externe Fehlerursache. Die meisten internen Fehlerursachen pendeln zwischen inaktivem und aktivem Zustand hin und her. Physikalische Fehlerursachen können nur Hardware-Komponenten direkt beeinflussen, während menschliche Fehler jede Komponente betreffen können.
2) Ein Fehlzustand kann **latent** oder erkannt sein. Ein Fehlzustand ist latent, wenn er noch nicht als solcher erkannt worden ist; ein Fehlzustand wird durch einen Erkennungsalgorithmus oder - mechanismus **erkannt**. Ein Fehlzustand kann verschwinden, bevor er erkannt ist. Ein Fehlzustand kann, und tut dies in der Regel auch, propagieren: durch die Propagation erzeugt ein Fehlzustand andere - neue - Fehlzustände. Während des Betriebs ist die Anwesenheit von aktiven Fehlerursachen nur durch die Erkennung von Fehlzuständen feststellbar.
3) Ein Ausfall tritt auf, wenn ein Fehlzustand die Schnittstelle System/Benutzer durchdringt und die Leistung, die das System erbringt, beeinträchtigt. Ein Komponentenausfall resultiert wiederum in einer Fehlerursache a) für das System, das die Komponente beinhaltet, und/oder b) für andere Komponenten, mit denen die Komponente zusammenarbeitet; die Ausfallarten der ausgefallenen Komponente werden dann Fehlerursachen für die Komponenten, mit denen sie zusammenarbeitet.

Durch diese Mechanismen kann die 'fundamentale Kette' vervollständigt werden:

$\cdots \rightarrow$ Ausfall $\rightarrow$ Fehlerursache $\rightarrow$ Fehlzustand $\rightarrow$ Ausfall $\rightarrow$ Fehlerursache $\rightarrow \cdots$

Wir wollen einige erläuternde Beispiele für die Fehlerpathologie geben:

- die Folge eines Programmierer*fehlers* ist eine (*inaktive*) *Fehlerursache* im Programm (fehlerhafte Anweisung oder Daten); bei der Aktivierung (Aufruf der Komponente, die die Fehlerursache enthält, und Verwendung der fehlerhaften Anweisung, Anweisungsfolge oder

Daten durch entsprechende Eingabedaten) wird die Fehlerursache *aktiv* und erzeugt einen Fehlzustand; wenn die fehlerhaften Daten die erbrachte Leistung beeinträchtigen (im Wert und/oder im Zeitverhalten der Erbringung), tritt ein *Ausfall* auf;

- ein Kurzschluß, der in einem integrierten Schaltkreis auftritt, ist ein *Ausfall* (in Hinblick auf die Leistungsspezifikation für den Schaltkreis); die Konsequenz ("stuck-at"-Fehler, Veränderung der Schaltkreisfunktion, usw.) ist eine *Fehlerursache*, die solange inaktiv sein wird, wie sie nicht aktiviert wird; der Prozeß setzt sich so fort wie im vorherigen Beispiel;
- eine elektromagnetische Störung von ausreichender Energie ist eine *Fehlerursache*; diese Fehlerursache kann
  a) direkt einen *Fehlzustand* erzeugen, z. B. durch elektromagnetische Interferenz mit den elektrischen Ladungen, die durch die Leitungen fließen,
  b) eine weitere (interne) Fehlerursache erzeugen; wenn sich z. B. die Störung auf die Speichereingabe in der Schreibposition durch die Veränderung einiger Bitwerte auswirkt, werden diese Fehler im weiteren Verlauf als Fehlerursachen im Speicher bleiben; diese Fehlerursachen werden wiederum inaktiv sein, bis diese bestimmten Speicherstellen gelesen werden; die Fehlzustand-Ausfall Folge von der externen transienten Fehlerursache über die interne Fehlerursache besteht immer noch auf dem elektronischen Niveau;
- eine falsche Maschinenbenutzung durch einen Bediener während des Betriebs eines Systems ist eine *Fehlerursache* (aus der Sicht des Systems); die daraus resultierenden veränderten Daten sind ein *Fehlzustand*; usw.
- der Fehler, den der Verfasser eines Wartungshandbuchs oder einer Bedienungsanleitung macht, kann zu einer Fehlerursache im entsprechenden Handbuch führen (fehlerhafte Anweisung), die solange inaktiv ist, wie diese Anweisung nicht aufgrund einer gegebenen Situation ausgeführt wird; usw.

Aus den oben gegebenen Beispielen sieht man leicht, daß die Inaktivität einer Fehlerursache sehr stark variieren kann, und zwar in Abhängigkeit von der Fehlerursache, von der Systembenutzung, usw.

Menschliche Fehlerursachen können entweder unbeabsichtigt oder beabsichtigt sein. Das obige Beispiel des Programmiererfehlers und seiner Konsequenzen kann folgendermaßen umformuliert werden: eine "logische

Bombe" ist von einem böswilligen Programmierer erzeugt; sie bleibt solange inaktiv, bis sie aktiviert wird (z. B. zu einem vorherbestimmten Datum); sie erzeugt dann einen Fehlzustand, der zu einem Speicherüberlauf oder der Verlangsamung der Programmausführung führen kann; als eine weitere Folge kann die Leistungserbringung unter der sogenannten Leistungsverweigerung leiden, einer besonderen Form des Ausfalls.

Diese Beispiele sind bewußt einfach gehalten. In der Wirklichkeit ist es normalerweise viel komplizierter. Dazu vier Beispiele:

a) eine bestimmte Fehlerursache in einer Komponente kann vielerlei Gründe haben; z. B. kann eine permanente Fehlerursache in einer physikalischen Komponente - z. B. Verharren auf Erdungspotential - resultieren von:
   - einem physikalischen Ausfall (z. B. verursacht durch einen Schwellwertwechsel),
   - einem Fehlzustand, der durch einen Entwurfsfehler verursacht ist — z. B. eine fehlerhafte Mikroinstruktion —, der sich von oben nach unten durch die Schichten durch auswirkt und einen Kurzschluß zwischen zwei Ausgängen des Schaltkreises hervorruft, der lange genug andauert, um einen Kurzschluß hervorzurufen, der die gleichen Folgen wie ein Schwellwertwechsel hat;
b) eine Fehlerursache einer bestimmten Klasse kann durch Fehlerpropagation eine Fehlerursache einer anderen Klasse erzeugen; z. B. könnte eine Fehlerursache, die während der Ausführung der o. g. Mikroinstruktion zu einem Fehlzustand geführt hat, eine transiente Fehlerursache sein;
c) einige Blickwinkel können — zumindest für eine gewisse Zeit — während des Propagationsprozesses weniger wichtig sein; was z. B. externe Fehlerursachen betrifft, die während der Ausführung von Software-Komponenten zu Eingabefehlern führen (und sie dadurch in ihrem sogenannten Ausnahme-Eingabebereich aufrufen [Cri 80]), so kann die Tatsache, daß die Fehlerursache physikalisch oder menschlich ist, für das Ausfallverhalten der Komponente unwichtig sein;
d) ein Ausfall ist oft die Folge der kombinierten Aktion verschiedener Fehlerursachen; dies trifft insbesondere zu, wenn manVertraulichkeitsprobleme betrachtet: eine Hintertür (z. B. ein Weg, um die Zugriffskontrolle zu umgehen), die absichtlich oder unabsichtlich in ein Rechensystem eingebaut ist, ist eine Entwurfsfehlerursache; diese Fehlerursache kann solange inaktiv sein, bis ein böswilliger Mensch sie benutzt, um in das System zu gelangen; das Einschalten des Eindringlings ist eine bewußte

Bedienungsfehlerursache; wenn der Eindringling eingeschaltet ist (was er/sie nicht sein sollte), kann er/sie absichtlich einen Fehlzustand erzeugen, z. B. einige Dateien verändern (Integritätsangriff); wenn diese Dateien von einem autorisierten Benutzer verwendet werden, ist die Leistung beeinträchtigt und ein Ausfall tritt ein.

Zwei weitere Kommentare, bezogen auf die Worte - oder Bezeichnungen - "Fchlcrursachc", "Fchlzustand" und "Ausfall":

a) ihre ausschließliche Verwendung in diesem Papier soll nicht die Verwendung von Worten, die eine spezielle Klasse von Mängeln kurz und unmißverständlich beschreiben, in besonderen Situationen ausschließen; dies trifft insbesondere zu auf Fehlerursachen (z. B. Defekt, Abweichung) und auf Ausfälle (z. B. Abbruch, Fehlfunktion, Leistungsverweigerung);
b) die Bedeutung, die den einzelnen Begriffen Fehlerursache, Fehlzustand und Ausfall zugewiesen wird, berücksichtigt z. T. auch den augenblicklichen Sprachgebrauch[12]: i) Fehler-Vermeidung, -Toleranz, und -Diagnose, ii) Fehlererkennung und -Korrektur, iii) Ausfallrate.

Schließlich soll darauf hingewiesen werden, daß die in diesem Abschnitt angegebenen Definitionen *syntaktisch* sind; entsprechend sind die Kriterien für die verschiedenen Klassifikationen betont worden und sind unserer Meinung nach wichtiger als die Klassen selber.

5. ZUVERLÄSSIGKEITSMITTEL

5.1 - Abhängigkeiten zwischen den Zuverlässigkeitsmitteln

Die Methoden, die in den grundlegenden Definitionen in Abschnitt 1 angeführt sind, sind in der Tat Ziele, die nicht ganz erreicht werden können, da alle entsprechenden Aktivitäten menschliche Aktivitäten und damit nicht perfekt sind. Diese Unzulänglichkeiten bringen *Abhängigkeiten* mit sich, die erklären, warum nur die *konzentrierte* Benutzung der oben genannten Methoden — vorzugsweise in jeder Stufe des Entwurfs-und Implementierungsprozesses — zu einem zuverlässigen Rechensystem führen

12 Wobei im Deutschen bei zusammengesetzten Begriffen die Unterscheidung zwischen Fehlerursache und Fehlzustand meist entfällt und in beiden Fällen 'Fehler' verwendet wird.

kann. Diese Abhängigkeiten können folgendermaßen beschrieben werden: trotz Fehlerverhinderung durch Anwendung von Entwurfsmethoden und Konstruktionsrichtlinien (unvollkommen, um nicht zu komplex zu sein) werden Fehler gemacht. Daher besteht Bedarf für die Fehlerbeseitigung. Die Fehlerbeseitigung ist aber auch wiederum unvollkommen, genauso wie die Standardkomponenten - Hardware wie Software - des Systems, daher die Bedeutung der Fehlervorhersage. Die steigende Abhängigkeit von Rechensystemen führt zur Forderung von Fehlertoleranz, die auch auf Konstruktionsregeln basiert. Das bringt uns wieder zur Fehlerbeseitigung, Fehlervorhersage, usw. Es muß angemerkt werden, daß dieser Prozeß sogar noch stärker rekursiv ist als es aus dem oben gesagten erscheint: bestehende Rechensysteme sind so kompliziert, daß zu ihrem Entwurf und ihrer Implementierung rechnergestützte Werkzeuge verwendet werden müssen, um kosteneffektiv zu bleiben (in einem weiten Sinn, einschließlich der Fähigkeit, innerhalb einer akzeptablen Zeit erfolgreich fertig zu sein). Diese Werkzeuge müssen ihrerseits auch wieder zuverlässig sein.

Die obige Argumentation zeigt die engen Zusammenhänge zwischen Fehlerbeseitigung und Fehlervorhersage und motiviert ihre Zusammenfassung unter dem einheitlichen Begriff *Validation*. Dies geschieht, obwohl Validation häufig eingeschränkt ist auf Fehlerbeseitigung und assoziiert ist zu einer Hauptaktivität bei der Fehlerbeseitigung, der Verifikation: z.B. in "V und V" (Validation und Verifikation) [Boe 79]; in einem solchen Fall ist die Unterscheidung so wie der Unterschied zwischen "das System richtig bauen" (entspricht der Verifikation) und "das richtige System bauen" (entspricht der Validation)[13]. Was hier vorgeschlagen wird, ist einfach eine Erweiterung dieses Konzepts: die Antwort auf die Frage "baue ich das richtige System?" (Fehlerbeseitigung) wird ergänzt durch "für wie lange wird es richtig sein?" (Fehlervorhersage)[14]. Außerdem ist die Fehlerbeseitigung normalerweise eng verbunden mit der Fehlerverhinderung und bildet mit ihr die **Fehlervermeidung**, d. h. "wie *baut man ein* fehlerfreies System". Neben der Hervorhebung der Notwendigkeit, die Prozeduren und Mechanismen der Fehlertoleranz zu validieren, ist es von großem Interesse, Fehlerbeseitigung und Fehlervorhersage als die beiden Bestandteile von ein und derselben Aktivität - der Validation - zu betrachten. Durch die obige Rekursion triff eine weitere Problematik auf: *die Validation der Validation*, oder wie erreiche ich

---

13 Es soll angemerkt werden, daß diese Zuordnung manchmal andersherum ist, wie z. B. im Bereich der Kommunikationsprotokolle (vgl. z. B. [Rud 85]).

14 Validation kommt von Validity = Gültigkeit, was zwei Bedeutungen hat:
- Gültigkeit zu einem bestimmten Moment, was in bezug steht zur Fehlerbeseitigung,
- Gültigkeit für eine bestimmte Zeit, was in bezug steht zur Fehlervorhersage.

Vertrauen in die Methoden und Werkzeuge, die benutzt wurden, um Vertrauen in das System zu erreichen. In diesem Zusammenhang ist der Begriff der Überdeckung hilfreich. **Überdeckung** ist ein Maß dafür, wie repräsentativ die Situation ist, der das System in der Validation ausgesetzt ist, verglichen mit der aktuellen/wahren Situation, der das System im Betrieb ausgesetzt ist[15]. Unvollkommene Überdeckung verstärkt die Beziehung zwischen Fehlerbeseitigung und Fehlervorhersage, da man sagen kann, daß die Notwendigkeit für die Fehlervorhersage mit geringerer Überdeckung der Fehlerbeseitigung steigt.

Im weiteren Verlauf dieses Kapitels betrachten wir nacheinander Fehlertoleranz, Fehlerbeseitigung und Fehlervorhersage; Fehlerverhinderung wird nicht weiter behandelt, da es eindeutig zur allgemeinen Systementwicklung gehört.

5.2- Fehlertoleranz

Fehlertoleranz [Avi 67] wird durch Fehlzustandsbehandlung und durch Fehlerursachenbehandlung [And 81] erreicht. **Fehlzustandsbehandlung** zielt ab auf die Beseitigung von Fehlern aus dem Rechenzustand, und zwar möglichst vor dem Eintreten eines Ausfalls; **Fehlerursachenbehandlung** zielt ab auf die Verhinderung, daß eine Fehlerursache aktiviert wird.

Fehlzustandbehandlung oder auch *Fehlerbehandlung* kann auf zwei Arten erreicht werden:

- **Fehlerbehebung**, indem ein fehlerfreier Zustand den fehlerhaften Zustand ersetzt; diese Ersetzung kann zwei Formen haben [And 81]:
  - **Rückwärtsfehlerbehebung**, bei der die Transformation des fehlerhaften Zustandes darin besteht, das System in einen Zustand zurückzuversetzen, in dem es vor dem Fehler schon einmal war. Dies setzt die Einrichtung von **Rücksetzpunkten** voraus. Dies sind Zeitpunkte in der Ausführung eines Prozesses, zu denen der dann aktuelle Zustand gespeichert wird, um später - falls erforderlich - wiederhergestellt werden zu können.

---

[15] Die Bedeutung der Überdeckung, wie sie hier definiert ist, ist sehr allgemein; sie kann dadurch stärker präzisiert werden, daß ihr Anwendungsgebiet angegeben wird, z. B.:
- Überdeckung des Software-Tests in Hinblick auf seinen Text, Kontrollgraphen, usw.
- Überdeckung des Schaltkreis-Tests im Hinblick auf das Fehlermodell,
- Überdeckung der Fehlertoleranz im Hinblick auf die Fehlerklassen,
- Überdeckung der Entwurfsannahmen im Hinblick auf die Realität.

- **Vorwärtsfehlerbehebung**, bei der versucht wird, einen neuen fehlerfreien Zustand zu finden, von dem aus das System wieder arbeiten kann (oft in einem reduzierten Umfang).

• **Fehlerkompensation**, bei der der fehlerhafte Zustand genügend Redundanz enthält, um die fehlerfreie Leistungserbringung trotz fehlerhaftem (internen) Zustand zu ermöglichen.

Die Fehlerbehebung setzt voraus, daß der fehlerhafte Zustand (schnellstens) als fehlerhaft identifiziert wird, bevor eine Zustandstransformation stattfindet; dafür wird die **Fehlererkennung** eingesetzt, und daher stammen auch die Begriffe *Fehlererkennung und -behebung*, die normalerweise benutzt werden. Die Verbindung von Datenverarbeitung und Fehlererkennung in einer Komponente führt zu dem Begriff der **selbstprüfenden Komponente**, entweder in Hardware [Car68, Wak 78, Nic 89] oder in Software [Yau 75, Lap 90a]; ein wichtiger Vorteil von selbstprüfenden Komponenten ist die Fähigkeit, eine klare Definition für *Fehlereingrenzungsbereiche* zu geben [Sie 82]. Wenn Fehlerkompensation in einem System durchgeführt wird, das aus selbstprüfenden Komponenten besteht, die aufgrund der Ausführung von der gleichen Task in Klassen eingeteilt sind, so ist die Zustandstransformation nichts anderes als der Wechsel innerhalb einer Klasse von einer ausgefallenen Komponente zu einer, die nicht ausgefallen ist. Daher der entsprechende Ansatz zur Fehlertoleranz: *Fehler-Erkennung und -Kompensation*[16]. Andererseits kann Kompensation selbst bei Abwesenheit von Fehlern durch **Fehlermaskierung** systematisch angewendet werden, (z. B. bei der Mehrheitsbewertung). Dies kann aber zu einer Reduzierung der Redundanz führen, die nicht bekannt ist. Daher beinhalten praktische Implementierungen der Maskierung im allgemeinen auch die Fehlererkennung, die dann *nach* der Zustandstransformation durchgeführt werden kann.

Rückwärts- und Vorwärts-Fehlerbehebung schließen sich nicht aus: Rückwärtsbehebung kann zunächst versucht werden; wenn der Fehler bestehen bleibt, kann anschließend Vorwärtsbehebung versucht werden. Bei der Vorwärtsbehebung ist es notwendig, den *Schaden zu bestimmen*, den ein erkannter Fehler angerichtet hat oder den Fehler angerichtet haben, bevor sie erkannt wurden; die Schadensermittlung kann - im allgemeinen - bei der Rückwärtsbehebung ignoriert werden, falls die Mechanismen zur Zustandstransformation nicht betroffen sind [And 81].

---

[16] Fehler-Erkennung und -Kompensation kann als ein Grenzfall der Fehler-Erkennung und -Behebung gesehen werden, bei dem die Behebung realisiert wird, indem der augenblickliche (fehlerhafte) Zustand des Systems benutzt wird, statt ihn durch einen fehlerfreien Zustand zu ersetzen.

Der Rechenzeit-Overhead für die Fehlerbehandlung hängt sehr stark von der gewählten Form der Fehlerbehandlung ab:

- bei der Fehlerbehebung wächst der Zeitbedarf, wenn ein Fehler aufgetreten ist; besonders bei der Rückwärtsbehebung hängt dieser von der Häufigkeit der Rücksetzpunkte ab;
- bei der Fehlerkompensation ist der Zeitbedarf nahezu unabhängig davon, ob ein Fehler auftritt oder nicht[17].

Die Dauer bei der Fehlerkompensation ist aufgrund der größeren Menge an (struktureller) Redundanz viel kürzer als die Dauer bei der Fehlerbehebung. Diese Bemerkung hat

a) große praktische Bedeutung, da die Entscheidung für die gewählte Fehlertoleranzstrategie in Abhängigkeit von der Zeitgranularität für den Benutzer gefällt wird;

b) eine Beziehung zwischen dem Overhead an Ausführungszeit und der strukturellen Redundanz eingeführt; allgemeiner, ein redundantes System zeigt immer redundantes Verhalten, was zumindest einigen Overhead an Ausführungszeit kostet; der Zeitmehrbedarf kann so klein sein, daß er für den Benutzer nicht wahrnehmbar ist, was besagt, daß die *Leistung* nicht redundant ist; eine extrem andere Form ist die 'Zeitredundanz' (redundantes *Verhalten*, das durch Wiederholung erreicht wird), die zumindest durch eine geringe, aber vorhandene strukturelle Redundanz initialisiert werden muß; grob gesagt, je mehr strukturelle Redundanz, desto weniger Zeitmehrbedarf.

Der erste Schritt bei der *Fehlerursachenbehandlung* ist die **Fehlerursachendiagnose**, die die Ursache(n) für die Fehler - Ort und Art - zu bestimmen hat. Danach kommen Aktionen, die verhindern sollen, daß die Fehlerursachen wieder aktiviert werden. Sie sind darauf ausgerichtet, die Fehlerursachen zu passivieren, d. h. **Fehlerursachen-Ausgliederung**. Dies geschieht dadurch, daß die Komponenten, die als fehlerhaft identifiziert wurden, von der weiteren Ausführung ausgeschlossen werden. Falls das System damit nicht mehr in der Lage ist, dieselbe Leistung zu erbringen wie vorher, kann eine *Rekonfiguration* stattfinden.

Falls zu erwarten ist, daß die Fehlerbehebung die Fehlerursache direkt beseitigt, oder wenn die Wahrscheinlichkeit des Wiederauftretens klein genug ist, so kann auf die Fehlerursachen-Ausgliederung verzichtet werden. Solange die Fehlerursachen-Ausgliederung nicht durchgeführt worden ist, wird die

---

17 In jedem Fall muß die Zeit, die für das Updaten der Datei mit dem Systemzustand benötigt wird, zum Zeitbedarf hinzugerechnet werden.

Fehlerursache als **weicher Fehler** betrachtet; ihre Ausführung hat zur Folge, daß die Fehlerursache als **harter Fehler** angesehen wird. Auf den ersten Blick können die Bedeutungen von weichen und harten Fehlerursachen als synonym zu den vorher eingeführten Begriffen der temporären und permanenten Fehlerursachen gesehen werden. In der Tat macht die Toleranz von temporären Fehlern eine Fehlerursachenbehandlung nicht erforderlich, wenn nicht eine permanente Fehlerursache durch den Propagationsprozeß hervorgerufen worden ist. In diesem Fall soll die Fehlerbehebung die Effekte der Fehlerursachen beseitigen, da die Fehlerursache selbst bereits verschwunden ist. Allerdings sind die Benennungen weicher und harter Fehler aus folgenden Gründen sinnvoll:

- es ist eine schwierige Aufgabe, einen permanenten Fehler von einem temporären Fehler zu unterscheiden, da a) ein temporärer Fehler nach einer gewissen Zeit verschwindet, normalerweise bevor eine Fehlerdiagnose durchgeführt wurde, und b) Fehlerursachen aus unterschiedlichen Klassen zu sehr ähnlichen Fehlzuständen führen können; daher drückt die Benennung von weichen und harten Fehlern in der Tat die Subjektivität aus, die mit diesen Schwierigkeiten verbunden ist, einschließlich der Tatsache, daß ein Fehler als ein weicher Fehler klassifiziert wird, wenn die Fehlerdiagnose erfolglos ist;
- diese Begriffe ermöglichen es, diffizile Fehlerursachen zu unterscheiden; kann man z. B. sagen, daß die inaktiven internen Fehlerursachen, die von Alpha-Teilchen verursacht werden (aufgrund der Ionisierung der Schaltkreisbauteile) oder von schweren Ionen im Weltraum auf Speicherelementen (im weitesten Sinne, einschließlich der Flip-Flops), *temporäre* Fehlerursachen sind? So eine inaktive Fehlerursache ist jedoch ein *weicher* Fehler.

Die vorangegangenen Definitionen beziehen sich sowohl auf physikalische Fehlerursachen als auch auf Entwurfsfehlerursachen; die Klasse von Fehlerursachen, die wirklich toleriert werden kann, hängt von der Fehlerursachen-Hypothese ab, die im Entwurfsprozeß betrachtet wird, und hängt damit von der *Unabhängigkeit* der Redundanzen bezogen auf die Fehlerursachen-Erstellung und -Aktivierung ab. Eine (oft benutzte) Methode, Fehlertoleranz zu erreichen, ist die Verwendung von mehreren Kanälen zur mehrfachen Berechnung. Wenn die Toleranz von physikalischen Fehlern vorgesehen ist, können die Kanäle identisch sein, basierend auf der Annahme, daß die Hardware-Komponenten unabhängig voneinander ausfallen. Eine derartige Vorgehensweise ist für die Toleranz von Entwurfsfehlerursachen nicht sinnvoll; hier müssen die Kanäle *identische Leistung* durch *getrennte*

Entwürfe und Implementierungen zur Verfügung stellen [Elm 72, Ran 75, Avi 78], z. B. durch **Entwurfsdiversität** [Avi 84].

Wenn man sich mit fehlertoleranten Systemen befaßt, begegnet man oft Situationen, die mehrfache Fehlerursachen und/oder Ausfälle betreffen. Betrachtet man ihre Ursachen, so kommt man zu folgender Unterscheidung:

- **unabhängige Fehlerursachen**, die unterschiedlichen Ursachen zuzuordnen sind,
- **verwandte Fehlerursachen**, die einer gemeinsamen Ursache zuzuordnen sind.

Beispiele für gemeinsame Ursachen sind (singuläre) Netzteile, Uhren, Spezifikationen usw. Bei Verwendung von Entwurfsdiversität können verwandte Fehlerursachen auch von Abhängigkeiten in den separaten Entwürfen und Implementierungen resultieren. Verwandte Fehlerursachen verursachen in der Regel **common-mode-Ausfälle**.

Wenn man fehlertolerante Systeme betrachtet, so ist der Begriff '**gleichzeitige Fehler**' nützlich und bezeichnet Fehler, die sich bei der gleichen Eingabe ergeben [Avi 86].

Die zeitliche Beziehung zwischen Mehrfach-Ausfällen führt zu folgender Unterscheidung:

- **Simultan-Ausfälle** sind Ausfälle, die innerhalb eines vordefinierten Zeitintervalls auftreten,
- **Folge-Ausfälle** sind Ausfälle, die nicht innerhalb desselben vordefinierten Zeitintervalls auftreten.

Obwohl eine mathematische Definition von 'simultan' das Zeitintervall praktisch zu Null setzt, kann das Zeitintervall in Wirklichkeit in Abhängigkeit von der Anwendung und der Zeit-Granularität des Benutzers recht groß sein. Daher benutzt man auch die Bezeichnung 'fast-gleichzeitige' Ausfälle [Tri 84]. In fehlertoleranten Systemen, die auf die Tolerierung von einem Fehler zu einem Zeitpunkt ausgelegt wurden, ist es normalerweise notwendig, sich von den Auswirkungen eines Fehlers zu erholen, bevor das System einen weiteren Fehler tolerieren kann. In diesem Zusammenhang ist das Zeitfenster, das die simultanen Fehler von den Folgefehlern trennt, das Zeitintervall, das für die Fehlerbehandlung erforderlich ist und während dem das System durch einen weiteren Fehler verletzbar ist.

Die Verhinderung von Fehlerpropagation auf die Funktionsfähigkeit von nicht-ausgefallenen Komponenten ist ein wichtiger Aspekt in der Koordinierung der Aktivitäten von verschiedenen Komponenten. Dies ist besonders wichtig, wenn eine Komponente ihre privaten Informationen an andere Komponenten übertragen muß. Typische Beispiele für solche *Einzel-Quellen-Informationen* sind lokale Sensordaten, der Wert der lokalen Uhr, die lokale Sicht des Status von anderen Komponenten, usw. In diesen Fällen müssen die nicht ausgefallenen Komponenten eine Übereinstimmung erreichen, wie sie die Information, die sie erhalten, jeweils konsistent zueinander verwenden wollen. Besondere Aufmerksamkeit ist diesem Problem im Bereich der verteilten Systeme gewidmet worden (siehe z. B. atomares Broadcast [Cri 85a], Uhrensynchronisation [Lam 85, Kop 87] oder Teilnehmerprotokolle [Cri 88]). Es ist wichtig zu erkennen, daß die unvermeidbare Anwesenheit von struktureller Redundanz in jedem fehlertoleranten System zum Problem der Übereinstimmung (Konsistenz) der redundanten Information führt. Zentrale fehlertolerante Systeme können Lösungen für das Übereinstimmungsproblem benutzen, die in einem "klassischen" verteilten System von Komponenten, die über Nachrichten miteinander kommunizieren, als zu kostspielig betrachtet werden (z. B. Zwischenstufen [Lal 86], mehrfache Stufen für interaktive Konsistenz [Fri 82]).

Die Kenntnis von Systemeigenschaften kann die notwendige Redundanz reduzieren, was zu einer sogenannten "billigen Fehlertoleranz" führt. Beispiele für solche Eigenschaften sind struktureller Natur: Codes zur Fehlererkennung und Fehlerkorrektur [Pet 72], robuste Datenstrukturen [Tay 80], Multiprozessoren und Rechnernetzwerke [Pra 86, Ren 86], algorithmen-basierte Fehlertoleranz [Hua 82]. Die tolerierten Fehler sind dann abhängig von den betreffenden Eigenschaften, da sie direkt in die Fehlerhypothesen eingehen.

Es ist wichtig, daß der Komponenten-Ausfall dem Benutzer signalisiert wird. Dies kann im Rahmen der *Ausnahmen* geschehen [Mel 77, Cri 80, And 81]. *Ausnahmebehandlungs*möglichkeiten, die in einigen Sprachen zur Verfügung gestellt werden, können einen guten Weg für die Implementierung der Fehlerbehebung darstellen, insbesondere von Vorwärtsbehebung[18].

Fehlertoleranz ist (ebenfalls) ein rekursives Konzept: die Mechanismen, die auf die Implementierung der Fehlertoleranz abzielen, müssen gegen Fehler geschützt werden, die sie bedrohen. Beispiele sind Voter-Verdopplung,

[18] Die Verwendung des Begriffs 'Ausnahme' muß aufgrund des Ursprungs, mit außergewöhnlichen Situationen fertig zu werden - nicht nur mit Fehlern - im Bereich der Fehlertoleranz vorsichtig verwendet werden: es könnte als ein Widerspruch erscheinen, daß Fehlertoleranz ein natürliches Attribut von Rechnersystemen ist, wenn man es von der ersten Designphase her betrachtet, und nicht ein 'außergewöhnliches' Attribut.

selbstprüfende Prüfer [Car 68], "stabiler" Speicher für Wiederanlauf-Programme und Rücksetzpunkt-Daten [Lam 81].

Fehlertoleranz ist nicht eingeschränkt auf zufällige Fehler. Schutz gegen Einbruch beinhaltet normalerweise Kryptographie [Den 82]. Einige Mechanismen der Fehlererkennung sind sowohl auf absichtliche als auch auf zufällige Fehler ausgerichtet (z. B. Speicherzugriffschutzverfahren), und es sind Verfahren vorgeschlagen worden, die sowohl Einbruch als auch physikalische Fehler [Fra 86, Rab 89] oder auch bösartig fehlerhafte Logik [Jos 88] tolerieren.

5.3- Fehlerbeseitigung

Fehlerbeseitigung besteht aus drei Schritten: Verifikation, Diagnose und Korrektur. **Verifikation** ist der Vorgang der Prüfung, ob ein System gewisse Eigenschaften erfüllt, die *Verifikationsbedingungen* genannt werden [Che 81]. Wenn es sie nicht erfüllt, müssen die anderen beiden Schritte vollzogen werden: Diagnose des Fehlers, der verhinderte, daß die Verifikationsbedingungen erfüllt wurden, und dann die Ausführung der notwendigen Korrekturen. Nach der Korrektur muß der Vorgang wiederholt werden, um zu überprüfen, daß die Fehlerbeseitigung keine ungewollten Folgen hatte; die in dieser Stufe durchgeführte Verifikation wird normalerweise **Regressionsverifikation** genannt. Die Verifikationsbedingungen können folgendermaßen formuliert werden:

- allgemeine Bedingungen, die für eine gegebene Klasse von Systemen zutreffen, die daher auch - relativ - unabhängig von der Spezifikation sind, z. B. Abwesenheit von Verklemmungen, Einhaltung von Entwurfs- und Realisierungsregeln;
- Bedingungen, die speziell für das betrachtete System sind und direkt aus der Spezifikation abgeleitet werden.

Die Verifikationstechniken können unterteilt werden je nachdem, ob sie die Ausführung des Systems beinhalten oder nicht. Verifikation ohne die Ausführung des Systems ist **statische Verifikation**, die folgendermaßen durchgeführt werden kann:

- für das System selbst a) in Form einer *statischen Analyse* (z. B. Inspektion oder Walk-Through [Mey 79], Datenflußanalyse [Ost 76], Komplexitätsanalyse [McC 76], Compilerprüfung) oder b) in Form eines *Korrektheitsbeweises* (induktive Assertionen [Hoa 69, Cra 87]);
- an einem Modell des Systems (z. B. Petri-Netze, endliche Automaten), was zu einer *Analyse des Verhaltens* führt [Dia 82].

Verifikation durch Ausführung des Systems nennt man **dynamische Verifikation**; die Eingaben für das System können entweder im Falle der **symbolischen Ausführung** symbolisch sein oder im Falle des Verifikationstestens konkrete Werte, was dann einfacher auch nur **Testen** genannt wird.

Ein System in bezug auf alle möglichen Eingaben erschöpfend zu testen, ist praktisch nicht machbar. Die Methoden zur Bestimmung der Testmuster können nach zwei Gesichtspunkten unterschieden werden: einerseits Kriterien für die Auswahl der Testeingaben, andererseits Erzeugung der Testeingaben.

Die *Kriterien* für die Auswahl der Testeingaben können wiederum nach drei Gesichtspunkten unterteilt werden:

- der Testgrund: die Überprüfung, ob ein System seine (funktionale) Spezifikation erfüllt, ist der **Konformitätstest**, während das Testen zur Fehleraufdeckung **fehlererkennendes Testen** genannt wird;
- das System-Modell: abhängig davon, ob die Funktion oder die Struktur des Systems getestet wird, spricht man von **funktionellem Testen** oder von **strukturellem Testen**;
- die Existenz eines Fehlermodells: wenn ein Fehlermodell existiert, wird **fehlerbasiertes Testen** durchgeführt [Mor 90], das auf die Aufdeckung bestimmter Klassen von Fehlern abzielt (z. B. stuck-at-Fehler in der Hardware-Herstellung [Rot 67], physikalische Fehler, die den Befehlssatz des Mikroprozessors beeinträchtigen [Tha 78], Entwurfsfehler in der Software [Goo 75, DeM 78, How 87]); wenn kein Fehlermodell vorliegt, können sich die Kriterien bei der Software auf die Pfad-Aktivierung [Rap 85, Nta 88], die Eingabebereichsgrenzen [Mye 79] usw. beziehen[19].

Die *Erzeugung* von Testeingaben kann deterministisch oder probabilistisch sein:

- beim **deterministischen Testen** werden die Testmuster aufgrund der angenommenen Kriterien vorherbestimmt,
- beim **statistischen Testen** werden die Testmuster aufgrund einer vorgegebenen Wahrscheinlichkeitsverteilung aus dem Eingabebereich ausgewählt; die Verteilung und die Anzahl der Eingabedaten sind durch die angenommenen Kriterien bestimmt [Dav 81, Dur 84].

---

[19] Die Möglichkeit, ein Fehlermodell zu definieren, ist eng gekoppelt mit den Stufen des Entwicklungsprozesses, die betrachtet werden: je weiter man im Entwicklungsprozeß vorangeschritten ist, desto größer ist die Möglichkeit, ein Fehlermodell zu definieren.

Die Verbindung von verschiedenen Blickwinkeln führt zu einer Reihe von Testverfahren, von denen einige hier dargestellt werden sollen:

- strukturelles Testen, das auf Hardware angewendet wird, bedeutet fehlererkennendes, strukturelles, fehlerbasiertes Testen, während die Anwendung auf Software im allgemeinen fehlererkennendes, strukturelles und nicht-fehlerbasiertes Testen bedeutet.
- *Mutationstest* für Software [DeM 78] ist fehlererkennendes, strukturelles, fehlerbasiertes und deterministisches Testen.

Die Beobachtung der Testausgaben und die Entscheidung, ob sie die Verifikationsbedingungen erfüllen oder nicht, ist als *Orakel*-Problem bekannt [Adr 82]. Die Verifikationsbedingungen können sich auf die gesamte Ausgabe beziehen oder auf eine kompakte Funktion dieser Daten (z. B. eine Systemsignatur, wenn in der Hardware auf physikalische Fehler getestet wird [Dav 86], oder ein "partielles Orakel", wenn die Software auf Entwurfsfehler getestet wird [Wey 82]). Wenn auf physikalische Fehler getestet wird, werden die Resultate - kompakt oder nicht - , die vom getesteten System zu einer gegebenen Eingabefolge erwartet werden, durch Simulation [Lev 86] oder von einem Referenzsystem ("goldenes System") bestimmt. Für Entwurfsfehler ist normalerweise die Spezifikation die Referenz; es kann aber auch ein Prototyp sein, oder eine andere Implementierung derselben Spezifikation wie im Fall der Entwurfsdiversität ("back-to-back" Testen, vgl. z. B. [Bis 88]).

Einige Verifikationsmethoden können gemeinsam verwendet werden, z. B. kann die symbolische Ausführung a) zur Bestimmung der Testmuster genutzt werden [Adr 82] oder b) als Korrektheitsbeweisverfahren [Car 78].

Da Verifikation während der gesamten Systementwicklung gemacht werden muß, lassen sich die oben genannten Techniken natürlich auf die verschiedenen Formen anwenden, die ein System während der Entwicklung einnimmt: Prototyp, Komponente, usw. Zu verifizieren, daß ein System nicht *mehr* machen kann als was spezifiziert worden ist, ist besonders wichtig im Hinblick auf die absichtlichen Fehler [Gas 88].

Ein System so zu entwerfen, daß die Verifikation erleichtert wird, heißt **verifikationsgünstiger Entwurf**. Dies ist insbesondere für Hardware im Hinblick auf physikalische Fehler üblich, wo die entsprechenden Techniken auch *testgünstiger Entwurf* genannt werden [Wil 83, McC 86].

Fehlerbeseitigung während der Betriebsphase eines Systems wird **Instandsetzung** genannt. Sie zielt ab auf die Erhaltung oder die Verbesserung der Fähigkeit des Systems, die der Spezifikation entsprechende Leistung zu erbringen[20]. Die Instandsetzung kann zwei Formen annehmen:

- **heilende** Instandsetzung, die darauf abzielt, Fehlerursachen zu beseitigen, die bereits einen oder mehrere Fehlzustände erzeugt haben und die gemeldet worden sind;
- **vorbeugende** Instandsetzung, die darauf abzielt, Fehlerursachen zu beseitigen, bevor sie Fehlzustände hervorgerufen haben; diese Fehlerursachen können sein
  - physikalische Fehlerursachen, die seit der letzten vorsorglichen Instandsetzung aufgetreten sind,
  - Entwurfsfehlerursachen, die in einem ähnlichen System bereits zu einem Fehlzustand geführt haben [Ada 84].

Diese Definitionen gelten sowohl für nicht fehlertolerante Systeme als auch für fehlertolerante Systeme, die on-line (ohne Unterbrechung der Leistungserbringung) oder off-line instandhaltbar sein können. Es ist schließlich noch bemerkenswert, daß die Grenze zwischen Instandsetzung und Fehlerbehandlung ziemlich willkürlich ist; insbesondere kann heilende Instandsetzung als ein - letztes - Mittel zum Erreichen von Fehlertoleranz betrachtet werden.

5.4- Fehlervorhersage

Fehlervorhersage wird vorgenommen, indem eine *Evaluation* des Systemverhaltens im Hinblick auf Fehlerauftreten und Fehleraktivierung durchgeführt wird. Die Evaluation hat zwei Aspekte:

- *nicht-probabilistisch*, z. B. Berechnung der minimalen Schnittmenge oder Pfadmenge eines Fehlerbaums, Durchführung einer Fehlerart- und Fehlereffektanalyse;
- *probabilistisch*, mit dem Ziel, die Übereinstimmung des Systems mit den Zuverlässigkeitszielvorgaben zu bestimmen, die in Form von Wahrscheinlichkeiten in Verbindung mit einigen Attributen der Zuverlässigkeit abgefaßt sind, welche dann als *Maße* für die Zuverlässigkeit definiert werden können.

---

[20] Die davon zu unterscheidenden Formen der Instandsetzung sind [Ram 84]:
- *adaptive Instandsetzung*, die das System den geänderten Umgebungsbedingungen anpaßt (z. B. Änderungen des Betriebssystems oder der System-Datenbasis);
- *perfektionierende Instandsetzung*, die die Systemfunktion verbessert aufgrund von Änderungsanforderungen des Benutzers oder des Entwicklers, und die die Beseitigung von Spezifikationsfehlern beinhalten kann.

Das Verhalten eines Systems wird von seinen Benutzern als ein Wechsel zwischen zwei Zuständen der erbrachten Leistung in Hinblick auf die Spezifikation gesehen:

- **korrekte Leistung**, wo die erbrachte Leistung mit der Spezifikation *übereinstimmt*[21];
- **inkorrekte Leistung**, wo die erbrachte Leistung nicht mit der Spezifikation *übereinstimmt*.

Ein Ausfall ist daher der Übergang von korrekter zu inkorrekter Leistung, und der Übergang von inkorrekter zu korrekter Leistung ist **Wiederherstellung**. Die Quantifizierung des Wechsels von korrekter und inkorrekter Leistungserbringung ermöglicht es, Funktionsfähigkeit und Verfügbarkeit als Maße für die Zuverlässigkeit zu definieren:

- **Funktionsfähigkeit** (oder Überlebenswahrscheinlichkeit): ein Maß für die *kontinuierliche* Erbringung von korrekter Leistung - oder gleichbedeutend für die Zeit bis zu einem Ausfall;
- **Verfügbarkeit**: ein Maß für die Erbringung der korrekten Leistung *in Hinblick auf den Wechsel* von korrekter und inkorrekter Leistung.

Ein drittes Maß, die **Instandhaltbarkeit**, wird normalerweise auch betrachtet. Sie kann als ein Maß für die Zeit vom letzten Ausfall bis zur Wiederherstellung definiert werden, oder gleichbedeutend als ein Maß für die kontinuierliche Erbringung von inkorrekter Leistung.

Sicherheit, als Maß betrachtet, kann als eine Erweiterung der Funktionsfähigkeit gesehen werden. Wenn der Zustand der korrekten Leistung mit dem Zustand der inkorrekten Leistung nach einem unkritischen Ausfall in einen sicheren Zustand zusammengefaßt wird (im Sinne von Freisein von katastrophalem Schaden, nicht von Gefahr), dann ist **Sicherheit** ein Maß für das kontinuierliche Sichersein, oder gleichbedeutend ein Maß für die Zeit bis zu einem kritischen Ausfall. Sicherheit kann somit interpretiert werden als Überlebenswahrscheinlichkeit in bezug auf kritische Ausfälle. Eine direkte Erweiterung der Verfügbarkeit, d. h. ein Maß von Sichersein im Hinblick auf den Wechsel von Sichersein und inkorrekter Leistung nach einem kritischen Ausfall, würde hingegen kein sinnvolles Maß ergeben. Wenn ein kritischer Ausfall aufgetreten ist, sind die Folgen im allgemeinen so schwerwiegend, daß eine Wiederherstellung der Leistung zumindest aus den folgenden zwei Gründen nicht die höchste Priorität hat:

---

[21] Wir schränken die Benutzung von 'korrekt' bewußt ein auf die Leistung, die ein System erbringt, und benutzen es nicht auf das System selbst: in unserer Betrachtung existieren fehlerfreie Systeme kaum; es gibt nur Systeme, die noch nicht ausgefallen sind.

- die Wiederherstellung der Leistung erfolgt erst nach der Reparatur (im weitesten Sinne des Begriffs, legale Aspekte inbegriffen) der Folgen des kritischen Ausfalls;
- die lange Zeitdauer, bevor das System wieder in Betrieb gehen darf (Untersuchungskommissionen usw.), würde zu unbedeutenden numerischen Werten führen.

Ein "hybrides" Überlebenswahrscheinlichkeits-Verfügbarkeits-Maß kann aber dennoch definiert werden: ein Maß für die korrekte Leistungserbringung im Hinblick auf den Wechsel von korrekter Leistung und inkorrekter Leistung nach einem unkritischen Ausfall. Dieses Maß ist von Interesse, da es in der Tat eine Quantifizierung der System-Verfügbarkeit *vor* einem kritischen Ausfall bedeutet und es damit ermöglicht, den sogenannten Trade-off zwischen Überlebenswahrscheinlichkeit (oder Verfügbarkeit) und Sicherheit zu quantifizieren.

Im Falle von Systemen, die viele Funktionen erfüllen, können verschiedene Leistungen und verschiedene Arten der Leistungserbringung unterschieden werden, von voller Leistung bis zu totaler Leistungsunterbrechung, und dies kann als eine graduelle Erbringung von immer weniger Leistung betrachtet werden. Erfüllungsbezogene Maße der Zuverlässigkeit für derartige Systeme werden normalerweise **Leistungsfähigkeit** genannt [Mey 78, Smi 88].

Die zwei wichtigsten Ansätze zur probabilistischen Fehlervorhersage, die darauf abzielen, quantifizierte Abschätzungen für die Zuverlässigkeitsmaße abzuleiten, sind die Modellierung und das Testen (die Evaluation). Diese Ansätze sind komplementär zueinander, da die Modellierung Daten über die zugrundeliegenden Prozesse benötigt (Ausfall-Prozeß, Instandhaltungsprozeß, Systembenutzungsprozeß usw.), die durch das Testen bereitgestellt werden können.

Wenn eine Evaluation durch *Modellierung* gemacht wird, unterscheiden sich die Ansätze entscheidend voneinander, je nachdem ob das System eine stabile Überlebenswahrscheinlichkeit haben soll oder eine wachsende Überlebenswahrscheinlichkeit, was folgendermaßen definiert werden kann [Lap 90b]:

- **stabile Überlebenswahrscheinlichkeit**: die Fähigkeit des Systems, korrekte Leistung zu erbringen, ist *erhalten* (stochastische Gleichheit von aufeinanderfolgenden Zeiten bis zum Ausfall);

- **Überlebenswahrscheinlichkeitswachstum**: die Fähigkeit des Systems, korrekte Leistung zu erbringen, *verbessert* sich (stochastisches Wachstum von aufeinanderfolgenden Zeiten bis zum Ausfall)[22].

Die praktische Interpretation der stabilen Überlebenswahrscheinlichkeit und der wachsenden Überlebenswahrscheinlichkeit ist folgendermaßen:

- stabile Überlebenswahrscheinlichkeit: nach einer Wiederherstellung ist das System identisch mit dem Zustand, in dem es bei der vorhergehenden Wiederherstellung war; dies korrespondiert zu der folgenden Situation:
  - im Falle von einem Hardware-Ausfall ist das ausgefallene Teil gegen ein anderes, identisches und nicht ausgefallenes ausgetauscht worden,
  - im Falle von einem Software-Ausfall ist das unveränderte System mit einem Eingabedatenmuster neu gestartet worden, das unterschiedlich ist zu dem, das zu dem Ausfall geführt hat;
- wachsende Überlebenswahrscheinlichkeit: die Fehlerursache, die zu dem Ausfall geführt hat, ist als ein Entwurfsfehler erkannt (in der Hardware oder in der Software) und beseitigt worden.

Die Evaluation der Zuverlässigkeit von Systemen mit stabiler Überlebenswahrscheinlichkeit verläuft normalerweise in zwei Hauptphasen:

- *Konstruktion* eines System-Modells aus den elementaren stochastischen Prozessen, die das Verhalten der System-Komponenten und ihre Interaktion modellieren;
- *Ausführung* des Modells, um die Ausdrücke und die Werte der Zuverlässigkeitsmaße des Systems zu erhalten.

Die Evaluation kann im Hinblick auf a) die physikalischen Fehler [Tri 84], b) die Entwurfsfehler [Lit 79, Arl 88], oder c) eine Kombination von beiden [Lap 84, Pig 88] durchgeführt werden. Die Zuverlässigkeit eines Systems hängt stark von seiner Umgebung ab, sei es im weitesten Sinne dieses Begriffs [Hec 87] oder sei es speziell bezogen auf die Belastung [Cas 81, Iye 82].

---

22 *Verminderung der Überlebenswahrscheinlichkeit* (die Fähigkeit des Systems, korrekte Leistung zu erbringen, nimmt ab, und es liegt eine stochastische Abnahme von aufeinanderfolgenden Zeiten bis zum Ausfall vor) ist sowohl theoretisch wie auch praktisch möglich, z. B. bei der Einführung von neuen Fehlern während einer Korrektur, deren Aktivierungswahrscheinlichkeit größer ist als die für die beseitigten Fehler. In einem solchen Fall kann man hoffen, daß die Verringerung nur für kurze Zeit gilt und daß die Überlebenswahrscheinlichkeit im großen und ganzen über einen längeren Beobachtungszeitraum gesehen wächst.

Viele Modelle für Systeme mit wachsender Überlebenswahrscheinlichkeit sind vorgeschlagen worden, eingeschränkt auf Hardware [Dua 64], auf Software, oder für beide gültig. Die meisten sind auf Software ausgerichtet, und sie zielen darauf ab, entweder die Überlebenswahrscheinlichkeit [Yam 85, Mil 86] oder die Zahl der (noch vorhandenen) Fehler [Goe 79, Toh 89] zu bestimmen; da diese Modelle darauf abzielen, die zukünftige Überlebenswahrscheinlichkeit aus Ausfalldaten, die in der Vergangenheit gesammelt wurden, vorherzusagen, hat man diesem Vorhersageproblem viel Aufmerksamkeit gewidmet [Lit 88].

Obwohl es nicht - in erster Linie - darauf abzielt, ein System zu verifizieren, kann das *Evaluationstesten* [Mil 87, Cho 87] mit den in Abschnitt 5.3 definierten Gesichtspunkten charakterisiert werden: Konformität, funktionell, nicht fehlerbasiert, statistisch. Ein wichtiger Punkt ist hierbei, daß das Eingabedatenprofil repräsentativ ist für das aktuelle Datenprofil während des Betriebs, daher auch der Name: **Nutzungstest**.

Wenn ein fehlertolerantes System evaluiert wird, hat die Überdeckung der Mechanismen der Fehlerbehandlung einen entscheidenden Einfluß [Bou 69, Arn 73]; seine Evaluation kann entweder mit Modellierung [Dug 89] oder durch Testen, das dann *Fehlerinjektion* heißt [Arl 89, Gun 89], durchgeführt werden.

6. Die Kenngrößen der Zuverlässigkeit

Die Kenngrößen sind in Kapitel 1 entsprechend den verschiedenen Eigenschaften definiert worden, die in Abhängigkeit von der geplanten Anwendung des betreffenden Rechensystems mehr oder weniger gewichtet werden können:

- Verfügbarkeit ist immer erforderlich, wenn auch abhängig von der Anwendung unterschiedlich stark;
- Überlebenswahrscheinlichkeit, Sicherheit, Vertraulichkeit können je nach Anwendung erforderlich sein oder nicht.

Eine weitere Eigenschaft, die als eine Voraussetzung für die Erfüllung von anderen Eigenschaften angesehen werden kann, ist die **Integrität**, d. h. die Bedingung, nicht beeinträchtigt zu sein, im weitesten Sinne dieses Wortes: a) für Daten oder Programme, und b) im Hinblick auf zufällige oder absichtliche Fehler.

Die unterschiedliche Gewichtung, die den Kenngrößen der Zuverlässigkeit gegeben werden kann, hat einen direkten Einfluß auf die passende Balance der Techniken, die im vorigen Kapitel besprochen worden sind und die angewandt werden sollen, um ein zuverlässiges System zu erhalten. Dies ist ein immer schwierigeres Problem, da einige der Attribute antagonistisch sind (z. B. Verfügbarkeit und Sicherheit, Verfügbarkeit und Vertraulichkeit) und daher der jeweilige Trade-off beachtet werden muß. Wenn man die drei wesentlichen Entwurfsdimensionen eines Rechensystems betrachtet, d. h. Kosten, Leistung und Zuverlässigkeit, verschlimmert sich das Problem noch weiter durch die Tatsache, daß die Zuverlässigkeitsdimension weniger verstanden wird als der Kosten-Leistung-Entwurfsraum [Sie 82].

Instandhaltbarkeit wurde in Abschnitt 5.4 als ein Maß für die Zuverlässigkeit definiert. Instandhaltbarkeit kann ebenso als eine Kenngröße der Zuverlässigkeit betrachtet werden, bezogen auf die Leichtigkeit, mit der Instandhaltungsarbeiten durchgeführt werden können.

Die Definition von Vertraulichkeit in Kapitel 1, die Verhinderung von nicht autorisiertem Zugriff und/oder nicht autorisierter Handhabung von Informationen, stammt von [ITS 90], wo Vertraulichkeit als die Kombination von Vertraulichkeit im engeren Sinne - der Verhinderung von nicht autorisierter Offenlegung von Informationen -, Integrität - der Verhinderung von nicht autorisierter Erweiterung oder Zerstörung von Information -, und Verfügbarkeit - der Verhinderung von nicht autorisierter Zurückhaltung von Information - eingeführt wird. Es ist wert zu bemerken, daß:

a) ein nicht autorisierter Zugriff oder eine nicht autorisierte Handhabung von Information von einem zufälligen oder einem absichtlichen Fehler herrühren kann, und daß - wie in Abschnitt 5.2 angemerkt - einige Mechanismen zum Schutz vor nicht autorisiertem Zugriff für beide Arten von Fehlern üblich sind;
b) im Hinblick auf absichtliche Fehler der Begriff der Autorisierung breit zu verstehen ist: eine autorisierte Person, die seine Autorisierung mißbraucht, verletzt in der Tat durch die illegale Handlung die Autorisierung, die ihr gegeben wurde.

Eine wichtige Eigenschaft eines fehlertoleranten Systems ist die Fähigkeit, die Benutzer aufgrund der Anwesenheit von Fehlererkennungsprozeduren mit Informationen zu versorgen, ob die erbrachte (funktionale) Leistung korrekt ist oder nicht. Solch eine Eigenschaft wird in [Fur 90] **Vertrauenswürdigkeit** genannt.

Die *Begutachtung*, ob ein System wirklich zuverlässig ist - gerechtfertigtes Vertrauen in die erbrachte Leistung - oder nicht, geht über die Validationstechniken hinaus, die in dem letzten Abschnitt behandelt wurden, zumindest aus den folgenden drei Gründen und Einschränkungen:

- Die Überdeckung von dem Entwurf oder den Validationsannahmen mit der Wirklichkeit (z. B. Relevanz der Kriterien, die für die Testdatenauswahl benutzt werden, für die wirklich auftretenden Fehler, Fehlerhypothesen im Entwurf von Fehlertoleranz-Mechanismen) mit völliger Gewißheit zu überprüfen, würde eine Kenntnis und eine Beherrschung der benutzten Technologie, der beabsichtigten Nutzung des Systems usw. bedingen, was bei weitem das übertrifft, was normalerweise erreichbar ist;
- eine Evaluation eines Systems entsprechend einiger Kenngrößen der Zuverlässigkeit im Hinblick auf einige Fehlerklassen durchzuführen, wird augenblicklich als nicht machbar oder nur unbedeutende Resultate ergebend angesehen: wahrscheinlichkeitstheoretische Grundlagen existieren nicht oder sind - derzeit - nicht weit akzeptiert; Beispiele sind Sicherheit im Hinblick auf zufällige Entwurfsfehler, Vertraulichkeit im Hinblick auf absichtliche Fehler;
- die Spezifikation, 'gegen' die die Validation durchgeführt wird, ist im allgemeinen auch nicht fehlerfrei - wie jedes System.

Unter den vielen Folgen dieser Sachlage wollen wir folgende erwähnen:

- die Bedeutung, die auf die Entwicklung und den Produktionsprozeß gelegt wird, wenn ein System begutachtet wird: welche Methoden und Techniken verwendet werden und wie sie eingesetzt werden; in einigen Fällen ist eine *Klasse* dem System zugewiesen entsprechend a) den Methoden und Techniken, die angewendet wurden, und b) einer Beurteilung seiner Benutzung[23];
- die Angabe in der Spezifikation von einigen fehlertoleranten Systemen, zusätzlich zu den probabilistischen Anforderungen an die Zuverlässigkeitsmaße, der Zahl von Fehlern, die toleriert werden sollen[24]; solch eine Spezifikation wäre nicht notwendig, wenn die oben erwähnten Einschränkungen überwunden werden könnten.

---

23 Zum Beispiel:
- Systeme werden entsprechend dem Vertraulichkeitsgesichtspunkt [DoD 85] eingeteilt von A1 ("verifizierter Entwurf") bis D ("minimaler Schutz");
- Software für zivile Flugzeuge ist klassifiziert [RTC 85] als Klasse 1, 2 oder 3 entsprechend der Kritikalität der Funktion, die von der Software wahrgenommen wird: kritisch, wichtig, oder nicht wichtig.

24 Solche Spezifikationen sind in Luftfahrt-Anwendungen typisch, z. B. in der Form der Verkettung von 'fail-operational' oder 'fail-safe' Anforderungen.

SCHLUßBEMERKUNG

In steigendem Maße werden komplexe Rechensysteme entwickelt oder realisiert, in deren Leistung großes Vertrauen gesetzt werden muß - ob nun ein Geldausgabegerät bedient, eine Satellitenumlaufbahn berechnet, ein Flugzeug oder ein Reaktor überwacht, oder die Vertraulichkeit einer sensitiven Datenbank gewährleistet werden soll. Unter unterschiedlichen Randbedingungen wird das Hauptinteresse auf unterschiedlichen Eigenschaften einer solchen Leistung liegen - z. B. die erreichte durchschnittliche Echtzeit-Antwortzeit, die Wahrscheinlichkeit, die geforderten Ergebnisse zu produzieren, die Fähigkeit, Ausfälle zu verhindern, die katastrophal für die Umgebung des Systems sein könnten, oder der Grad, zu dem beliebige Einbrüche verhindert werden können. Der Begriff der Zuverlässigkeit stellt ein adäquates Mittel zur Verfügung, um diese unterschiedlichen Belange innerhalb eines einheitlichen Konzeptrahmens zu behandeln. Zuverlässigkeit schließt somit als Spezialfälle solche Eigenschaften wie Überlebenswahrscheinlichkeit, Verfügbarkeit, Sicherheit und Vertraulichkeit mit ein. Sie gibt auch die Möglichkeit, das Problem anzusprechen, daß das, was ein Benutzer normalerweise von einem System braucht, eine *ausgewogene Balance* dieser Eigenschaften ist.

GLOSSAR

Warnung: Das Glossar soll nur als eine Hilfe beim Lesen dieses Dokuments dienen. Es soll nicht unabhängig von diesem Dokument verwendet werden. Insbesondere stellt es nicht eine Definition der Begriffe dar, wie sie normalerweise in Normen zu finden ist.

Absichtliche(r) Fehler (ursache).................................... Fehler, der absichtlich erzeugt wurde.

Aktive Fehler(ursache).............. Fehlerursache, die durch ihre Aktivierung einen Fehlzustand erzeugt hat.

Atomares System........................ System, dessen interne Struktur nicht erkennbar oder nicht von Interesse und daher ignorierbar ist.

Ausfall .... Abweichung der erbrachten Leistung von der in der System-Spezifikation geforderten Leistung. Übergang von korrekter Leistungserbringung zu fehlerhafter Leistungserbringung.

Ausfallsicheres System .... System, dessen Ausfälle in der Regel nur unkritische Ausfälle sind.

Ausgliederung (Fehlerursachen~) .... Maßnahmen zur Verhinderung, daß eine Fehlerursache aktiviert werden kann.

Auslassungsausfall .... Ausfall, so daß keine Leistung erbracht wird.

Beeinträchtigung der Zuverlässigkeit .... Unerwünschte, aber nicht unerwartete Störfaktoren, die Unzuverlässigkeit verursachen oder daraus resultieren. Fehlerursachen, Fehlzustände, Ausfälle.

Behandlung (Fehler~) .... Maßnahmen, die unternommen werden, um zu verhindern, daß ein Fehler reaktiviert wird, z. B. Beseitigung des Fehlers.

Behebung (Fehler~) .... Form der Fehlerbehandlung, bei der ein fehlerhafter Zustand durch einen fehlerfreien Zustand ersetzt wird.

Benutzer .... Ein anderes System (physikalisch, menschlich), das mit dem betrachteten System zusammenwirkt.

Benutzungsfehler(ursache) .... Fehler, der während der Nutzung des Systems auftritt.

Beseitigung (Fehler~) .... Methoden und Techniken, um die Anwesenheit (Zahl, Schwere) der Fehler zu reduzieren.

Bösartig fehlerhafte Logik .... Absichtlicher Entwurfsfehler.

Common-mode-Ausfälle .... Ausfälle, die auf die gleiche Fehlerursache zurückzuführen sind.

Deterministisches Testen .... Art des Testens, bei dem die Testdaten durch ein selektives Verfahren vorbestimmt sind.

Diagnose (Fehler~) .... Die Bestimmung eines Fehlers in bezug auf seinen Ort und seine Art.

Dynamische Verifikation .... Verifikation, die die Ausführung des Systems beinhaltet.

Echtzeit-Funktion........................Funktion, die innerhalb eines von der Umgebung vorgegebenen endlichen Zeitintervalles erfüllt werden muß.

Echtzeit-Leistung........................Leistung, die innerhalb eines von der Umgebung vorgegebenen endlichen Zeitintervalles erbracht werden muß.

Echtzeit-System..........................System, das mindestens eine Echtzeit-Funktion erfüllt oder mindestens eine Echtzeit-Leistung erbringt.

Eindringen / Einbruch...............Absichtlicher externer Bedienungsfehler. Beabsichtigte Herbeiführung eines Fehlers während des Betriebs.

Entwurfsdiversität......................Ein Verfahren bei der Erstellung eines Systems, das die gleiche Leistung von getrennten Entwürfen und Implementierungen ergibt.

Entwurfsfehler...........................Vom Menschen beim Entwurf verursachte interne Fehlerursache.

Ergebnisbezogener Ausfall........Ausfall, so daß das Ergebnis der erbrachten Leistung nicht mit der Spezifikation übereinstimmt.

Erkannter Fehler........................Fehler, der als solcher durch einen Erkennungsalgorithmus oder -mechanismus erkannt ist.

Erkennung (Fehler~).................Die Handlung, festzustellen, daß ein Systemzustand fehlerhaft ist. Der Vorgang des Erkennens eines Fehlers.

Externe Fehler(ursache)............Fehlerursache, die auf Umgebungseinflüssen oder externen Eingriffen beruht.

Fail-safe System.........................siehe Ausfallsicheres System

Fail-silent System.......................System, dessen Ausfälle in der Regel nur Total-Ausfälle sind.

Fail-stop System.........................System, dessen Ausfälle in der Regel nur Ausfälle sind, die zum Systemstillstand führen.

Fehler / Fehlerursache...............Anerkannte oder hypothetische Ursache für einen Fehler. Fehlerursache, die vermieden oder toleriert werden sollte. Auswirkung auf das betrachtete System durch den Ausfall eines anderen Systems, das mit dem System zusammengewirkt hat oder zusammenwirkt.

Fehler / Fehlzustand .... Teil des Systemzustandes, der dafür verantwortlich ist, daß ein Ausfall auftritt. Offenbarung einer Fehlerursache im System.

Fehlerbasiertes Testen .... Testen, das darauf abzielt, bestimmte Klassen von Fehlern aufzudecken.

Fehlererkennendes Testen .... Testen mit dem Ziel, Fehler aufzudecken.

Fehlerhafte Leistung .... Erbrachte Leistung, die nicht in Übereinstimmung mit der System-Spezifikation ist.

Fehlerursache .... siehe Fehler / Fehlerursache

Fehlerursache aufgrund menschlicher Einflüsse .... Fehler, der seine Ursache in der menschlichen Unzulänglichkeit hat.

Fehlzustand .... siehe Fehler / Fehlzustand

Folgeausfälle .... Ausfälle, die nicht innerhalb eines vordefinierten Zeitintervalls auftreten.

Funktion (System~) .... Wozu ein System gedacht ist.

Funktionsfähigkeit .... siehe Überlebenswahrscheinlichkeit

Funktionelles Testen .... Art des Testens, bei der die Testdaten aufgrund der Funktion des Systems ausgewählt werden

Gleichzeitige Fehler .... Fehler, die bei der gleichen Eingabe auftreten.

Harter Fehler .... Fehler, für den eine Fehlerursachen-Ausgliederung vorgenommen wird.

Heilende Instandsetzung .... Beseitigung von Fehlerursachen, die bereits Fehlzustände hervorgerufen haben.

Inaktiver Fehler .... Interne Fehlerursache, die im Rechenprozeß noch nicht aktiviert wurde.

Inkonsistenter Ausfall .... Ausfall, der nicht von allen Systembenutzern in der gleichen Form beobachtet wird.

Instandhaltbarkeit .... Maß für die Zeit vom letzten aufgetretenen Ausfall bis zur Wiederherstellung.

Instandsetzung .... Instandhaltung, die darauf abzielt, Fehlerursachen zu beseitigen, die Fehler hervorgerufen haben, die gemeldet worden sind. Erhaltung oder Verbesserung der Fähigkeit eines

	Systems, während seiner aktiven Lebenszeit die Funktion in Übereinstimmung mit der Spezifikation zu erbringen. Fehlerbeseitigung während der aktiven Lebenszeit eines Systems.
Integrität	Unversehrtheit.
Intermittierender Fehler	Kurzzeitiger interner Fehler. Fehler, dessen Aktivierungsbedingung nicht reproduziert werden kann oder die selten genug auftritt.
Interner Fehler	Fehler innerhalb eines Systems.
Kenngrößen der Zuverlässigkeit	Kenngrößen, die es ermöglichen, die Qualität des Systems zu beurteilen, die sich aus den Beeinträchtigungen und den dagegen eingesetzten Mitteln ergibt. Funktionsfähigkeit,Verfügbarkeit, Instandhaltbarkeit, Sicherheit, Vertraulichkeit, Vertrauenswürdigkeit.
Kompensation (Fehler~)	Art der Fehlerbehandlung, wenn der fehlerhafte Zustand genügend Redundanz beinhaltet, um die korrekte Funktion zu ermöglichen.
Komponente (System~)	Ein anderes System.
Konformitätstest	Test, um zu prüfen, ob das System die Spezifikation erfüllt.
Konsistenter Ausfall	Ausfall, der von allen Systembenutzern in der gleichen Form beobachtet wird.
Korrekte Leistung	Leistung, die in Übereinstimmung mit der Spezifikation erbracht wird.
Kritikalität	Grad der Auswirkungen bei den Ausfallarten.
kritischer Ausfall	Ausfall, dessen Folgen erheblich größer sind als der Nutzen bei korrekter Funktion.
Latenter Fehler	Fehler, der noch nicht erkannt ist.
Leistung	Systemverhalten, wie es vom Systembenutzer beobachtet wird.
Leistungsfähigkeit	Leistungsbezogenes Maß für die Zuverlässigkeit.

Mängel der Zuverlässigkeit.......Unerwünschte, aber nicht unerwartete Störfaktoren, die Unzuverlässigkeit verursachen oder daraus resultieren. Fehlerursachen, Fehler, Ausfälle.

Maskierung (Fehler~)...............Ergebnis der Anwendung von systematischer Fehlerkompensation, auch bei Abwesenheit von Fehlern.

Menschliche Fehler...................siehe Fehlerursache aufgrund menschlicher Einflüsse.

Nutzungstest...............................Test zur Bewertung der System-Zuverlässigkeit mit einem Eingabe-Profil, das den Nutzungsbedingungen entspricht.

Permanenter Fehler..................Fehler, der seine Passivierung erforderlich macht. Fehler, der nicht flüchtig ist.

Physikalische Fehler(ursache)...Fehler aufgrund eines physikalischen Phänomens.

Regressionsverifikation............Verifikation, die nach einer Korrektur durchgeführt wird, um zu überprüfen, daß die Korrektur keine ungewollten Auswirkungen hat.

Rücksetzpunkt............................Zeitpunkt bei der Ausführung eines Prozesses, für den der dann gültige Zustand später wieder hergestellt werden kann.

Rückwärtsfehlerbehebung..........Art der Fehlerbehebung, bei der der fehlerhafte Zustand in einen bereits vorher eingenommenen Zustand übergeführt wird.

Schweregrad (Ausfall-~)...........Grad der Auswirkungen eines Ausfalls auf die Umgebung.

Selbstprüfende Komponente......Komponente, die aus Fehlererkennungsmechanismen verbunden mit dem funktionellen Teil besteht.

Sicherheit....................................Zuverlässigkeit mit Bezug auf das Nichtauftreten von kritischen Ausfällen. Maß für die kontinuierliche Bereitstellung von korrekter Leistung oder inkorrekter Leistung nach einem gutartigen Ausfall. Maß für die Zeit bis zu einem kritischen Ausfall.

Simultan-Ausfälle........................Ausfälle, die innerhalb eines vordefinierten Zeitintervalls auftreten.

Spezifikation (System~).............Beschreibung der System-Anforderungen, auf die man sich geeinigt hat.

Stabile Überlebenswahrscheinlichkeit...Die Fähigkeit eines Systems, korrekte Leistung zu erbringen, ist stabil. Stochastische Gleichheit von aufeinander folgenden Ausfallzeiten.

Statische Verifikation................Verifikation, die ohne Ausführung des Systems durchgeführt wird.

Statistisches Testen....................Art des Testens, bei der die Testdaten gemäß einer definierten statistischen Verteilungsfunktion aus dem Eingabebereich gewählt werden.

Stillstand-Ausfall........................System-Ausfall, bei der eine Aktivität, falls überhaupt noch vorhanden, nicht mehr vom Benutzer beobachtbar ist und eine Leistung von konstantem Wert erbracht wird.

Struktur (System~).....................Was ein System das machen läßt, was es macht.

Strukturelles Testen..................Art des Testens, bei der die Testdaten gemäß Kriterien aufgrund der Systemstruktur gewählt werden.

Symbolische Ausführung...........Dynamische Verifikation mit symbolischen Eingabewerten.

System..Einheit, die mit anderen Einheiten zusammengearbeitet hat, zusammenarbeitet oder zusammenarbeiten kann. Menge von Komponenten, die miteinander verbunden sind, um zusammenzuarbeiten.

Temporärer Fehler....................Fehler, der nur eine beschränkte Zeit existiert.

Testen..Dynamische Verifikation mit konkreten Eingabewerten.

Toleranz (Fehler~).....................Methoden und Techniken, die darauf abzielen, trotz Fehlern eine Leistung zur Verfügung zu stellen, die mit der Spezifikation übereinstimmt.

Total-Ausfall..............................Totaler Ausfall des Systems.

Transienter Fehler .... Temporärer physikalischer externer Fehler.

Überdeckung .... Maß für die Güte der Übereinstimmung zwischen Situationen, denen ein System in der Validation und in der realen Benutzung ausgesetzt ist.

Überlebenswahrscheinlichkeit ... Zuverlässigkeit mit Bezug auf die Kontinuität der Leistung. Maß der kontinuierlichen korrekten Leistungserbringung. Maß für die Zeit bis zum Ausfall.

Überlebenswahrscheinlichkeitswachstum
.... Die Fähigkeit des Systems, korrekte Leistung zu erbringen, ist verbessert. Stochastischer Zuwachs bei aufeinanderfolgenden Zeiten bis zum Ausfall.

Umgebung (System~) .... Die anderen Systeme, die mit dem betrachteten System in Verbindung stehen.

Unabhängige Fehler(ursachen) .. Fehler, die unterschiedliche Ursachen haben.

Unkritischer Ausfall .... Ausfall, dessen Kosten in derselben Größenordnung liegen wie der Nutzen bei korrekter Funktion.

Validation .... Methoden und Techniken, die dazu dienen, daß Vertrauen in die Fähigkeit des Systems, Leistung gemäß der Spezifikation zu erbringen, erreicht wird. Fehlerbeseitigung und Fehlervorhersage.

Verfügbarkeit .... Zuverlässigkeit in bezug auf das Bereitsein zur Benutzung

Verhalten (System~) .... Was ein System macht.

Verhinderung (Fehler~) .... Methoden und Techniken mit dem Ziel, das Auftreten oder die Einführung von Fehlern / Fehlerursachen zu verhindern.

Verifikation .... Der Vorgang des Feststellens, ob ein System den Eigenschaften genügt (den Verifikation-Bedingungen), die - a) allgemein, unabhängig von der Spezifikation, oder - b) speziell, abgeleitet von der Spezifikation sind.

Verifikationsgünstiger Entwurf..Entwurf aus der Sicht der Verifikation. Methoden und Techniken, die beim Entwurf eines Systems angewandt werden und die Verifikation unterstützen.

Vermeidung (Fehler~)...............Methoden und Techniken, die darauf abzielen, ein fehlerfreies System zu erstellen. Fehlerverhinderung und Fehlerbeseitigung.

Versagen....................................siehe Ausfall.

Vertrauenswürdigkeit................Fähigkeit eines Systems, den Benutzer mit Informationen über die Leistungskorrektheit zu versorgen.

Vertraulichkeit...........................Zuverlässigkeit mit Bezug auf die Verhinderung von nicht autorisiertem Zugriff und/oder Behandlung von Informationen.

Verwandte Fehler(ursachen).....Fehler, denen eine gemeinsame Ursache zugeschrieben wird.

Vorbeugende Instandsetzung.....Korrigierende Instandhaltung mit dem Ziel, Fehler zu beseitigen, bevor sie aktiviert werden.

Vorhersage (Fehler~)................Methoden und Techniken, die dazu dienen, die aktuelle Zahl, das zukünftige Auftreten und die Folgen von Fehlern abzuschätzen.

Vorwärtsfehlerbehebung...........Art der Fehlerbehebung, bei der ein fehlerfreier Zustand gesucht wird.

Weicher Fehler...........................Fehler, für den keine Passivierung vorgenommen wird.

Werte-Ausfall.............................siehe Ergebnisbezogener Ausfall.

Wiederherstellung (Leistungs~) Übergang von fehlerhafter zu korrekter Leistungserbringung.

Willkürlicher Ausfall................siehe Ausfall.

Zeitbezogener Ausfall, Zeit-Ausfall................................Ausfall, so daß das Zeitverhalten der Leistungserbringung nicht mit der Spezifikation übereinstimmt.

Zufällige Fehler(ursache)..........Zufällig auftretende oder erzeugte Fehlerursache, unabsichtlicher Fehler.

Zustand (System~)......................Beschaffenheit einer Einheit mit Hinblick auf eine Menge von Randbedingungen.

Zuverlässigkeit..........................Vertrauenswürdigkeit eines Rechensystems, so daß Vertrauen in die erbrachte Leistung gesetzt werden kann.

Zuverlässigkeitsmittel................Methoden und Techniken, um a) ein System mit der Fähigkeit auszustatten, eine Leistung zu erbringen, in die Vertrauen gesetzt werden kann, und b) Vertrauen in diese Fähigkeit zu erreichen.

Zuverlässigkeitsverfahren.........Methoden und Techniken, die das System in die Lage versetzen, eine Leistung entsprechend der Spezifikation zu erbringen. Fehler-Vermeidung und Fehler-Toleranz.

Querverweis Deutsch - Englisch

Absichtlicher Fehler..................................Intentional fault
Aktive Fehlerursache................................Active fault
Atomares System.......................................Atomic system
Ausfall..Failure
Ausfallsicheres System..............................Fail-safe system
Ausgliederung (Fehler~)............................Passivation (fault ~)
Auslassungsausfall....................................Omission failure
Beeinträchtigung der Zuverlässigkeit ....Impairments to dependability
Behandlung (Fehlerursachen~)....................Treatment (fault~)
Behandlung (Fehlzustands~).......................Processing (error ~)
Behebung (Fehler~)...................................Recovery (error ~)
Benutzer...User (system ~)
Benutzungsfehler(ursache)..........................Operational fault
Beseitigung (Fehler~) ....Removal (fault ~)
Bösartig fehlerhafte Logik..........................Malicious logic
Common-mode-Ausfälle ....Common-mode failures
Deterministisches Testen............................Deterministic testing
Diagnose (Fehler~)....................................Diagnosis (fault ~)
Dynamische Verifikation............................Dynamic verification
Echtzeit-Funktion......................................Real-time function
Echtzeit-Leistung ....Real-time service
Echtzeit-System...Real-time system

Eindringen / Einbruch .... Intrusion
Entwurfsdiversität .... Design diversity
Entwurfsfehler.. Design fault
Ergebnisbezogener Ausfall......................... Value failure
Erkannter Fehler....................................... Detected error
Erkennung (Fehler~) .... Detection (error ~)
Externe Fehlerursache .... External fault
Fail-safe-System.. Fail-safe system
Fail-silent-System...................................... Fail-silent system
Fail-stop-System.. Fail-stop system
Fehler / Fehlerursache .... Fault
Fehler / Fehlzustand .... Error
Fehlerbasiertes Testen............................... Fault-based testing
Fehlererkennendes Testen .... Fault-finding testing
Fehlerhafte Leistung.................................. Incorrect service
Fehlerursache .... Fault
Fehlerursache aufgrund menschlicher
Einflüsse .... Human made fault
Fehlzustand.. Error
Folgeausfälle.. Sequential failures
Funktion (System~) .... Function (system ~)
Funktionsfähigkeit.................................... Reliability
Funktionelles Testen................................. Functional test
Gleichzeitige Fehler.................................. Coincident errors
Harter Fehler.. Hard fault, solid fault
Heilende Instandsetzung .... Curative maintenance
Inaktiver Fehler.. Dormant fault
Inkonsistenter Ausfall .... Inconsistent failure
Instandhaltbarkeit...................................... Maintainability
Instandsetzung.. Corrective maintenance
Integrität .... Integrity.
Intermittierender Fehler .... Intermittent fault
Interner Fehler .... Internal fault
Kenngrößen der Zuverlässigkeit .... Attributes of dependability
Kompensation (Fehler~)............................ Compensation (error ~)
Komponente (System~) .... Component (system ~)
Konformitätstest....................................... Conformance testing
Konsistenter Ausfall .... Consistent failure
Korrekte Leistung..................................... Correct service

Kritikalität..Criticality (system ~)
Kritischer Ausfall......................................Catastrophic failure
Latenter Fehler..Latent error
Leistung ....Service
Leistungsfähigkeit......................................Performability
Maskierung (Fehler~)................................Masking (fault ~)
Menschlicher Fehler...................................Human-made fault
Nutzungstest ....Operational testing
Permanenter Fehler...................................Permanent fault, solid fault, hard fault
Physikalische Fehlerursache ....Physical fault
Regressionsverifikation...............................Regression verification
Rücksetzpunkt...Recovery point
Rückwärtsfehlerbehebung ....Backward recovery
Schweregrad (Ausfall~).............................Severity (failure ~)
Selbstprüfende Komponente ....Self-checking component
Sicherheit ....Safety
Simultan-Ausfälle.......................................Simultaneous failures
Spezifikation (System~)............................Specification (system ~)
Stabile Überlebenswahrscheinlichkeit ....Stable reliability
Statische Verifikation................................Static verification
Statistisches Testen ....Statistical testing, random testing
Stillstand-Ausfall..Stopping failure
Struktur (System~) ....Structure (system ~)
Strukturelles Testen...................................Structural testing
Symbolische Ausführung............................Symbolic execution
System ....System
Temporärer Fehler....................................Temporary fault
Testen..Testing
Toleranz (Fehler~)....................................Tolerance (fault ~)
Total-Ausfall ....Crash failure
Transienter Fehler ....Transient fault
Überdeckung ....Coverage
Überlebenswahrscheinlichkeit.....................Reliability
Überlebenswahrscheinlichkeitswachstum......Reliability growth
Umgebung (System~)................................Environment (system ~)
Unabhängige Fehler(ursachen)...................Independent faults
Unkritischer Ausfall ....Benign failure
Validation...Validation

Verfügbarkeit .... Availability
Verhalten (System~) .... Behavior (system ~)
Verhinderung (Fehler~) .... Prevention (fault ~)
Verifikation .... Verification
Verifikationsgünstiger Entwurf .... Design for verifiability
Vermeidung (Fehler~) .... Avoidance (fault ~)
Versagen .... Failure
Vertrauenswürdigkeit .... Trustability
Vertraulichkeit .... Security
Verwandte Fehler(ursachen) .... Related faults
Vorbeugende Instandsetzung .... Preventive maintenance
Vorhersage (Fehler~) .... Forecasting (fault ~)
Vorwärtsfehlerbehebung .... Forward recovery
Weicher Fehler .... Soft fault
Werte-Ausfall .... Value failure
Wiederherstellung (Leistungs~) .... Restoration (service ~)
Willkürlicher Ausfall .... Arbitrary failure
Zeit-Ausfall .... Timing failure
Zeitbezogener Ausfall .... Timing failure
Zufällige Fehlerursache .... Accidental fault
Zustand (System~) .... State (system ~)
Zuverlässigkeit .... Dependability
Zuverlässigkeitsmittel .... Means for dependability
Zuverlässigkeitsverfahren .... Procurement of dependability

GARANZIA DI FUNZIONAMENTO:

CONCETTI BASE E

TERMINOLOGIA

INTRODUZIONE

Questo documento ha lo scopo di fornire definizioni informali ma precise che caratterizzano i vari attributi della garanzia di funzionamento dei sistemi di elaborazione. E' un contributo al lavoro intrapreso nella comunita' scientifica e tecnica di "Elaborazione Affidabile e Tollerante il Guasto" (Reliable and Fault Tolerant Computing) [Avi 67, Jes 77, Mel 77, Avi 78, Ran 78, Car 79, And 81, FTC 82, Sie 82, Cri 85a, Lap 85, Avi 86, Lap 89] per proporre definizioni chiare e largamente accettabili per alcuni concetti base.

La garanzia di funzionamento e' dapprima introdotta come concetto globale che include gli attributi usuali di affidabilita', disponibilita', sicurezza di funzionamento e sicurezza-confidenzialita'. Le definizioni di base date nel primo capitolo vengono poi commentate, e completate da definizioni aggiuntive, nei capitoli successivi. Un glossario e' dato in appendice; esso ricapitola le definizioni date nell'intero documento. La presentazione e' stata strutturata in modo da evitare riferimenti "in avanti". Vengono usati caratteri in grassetto quando un termine viene definito, caratteri in corsivo per focalizzare l'attenzione del lettore. Le linee che hanno guidato questa presentazione possono essere riassunte come segue:

- ricerca di un numero ridotto di concetti capaci di esprimere gli attributi della garanzia di funzionamento;
- dove possibile, uso dei termini che sono identici, o il piu' possibile vicini a quelli generalmente usati; come regola, un termine che non e' stato definito mantiene il suo significato ordinario (cosi' come fornito da un qualsiasi dizionario);
- enfasi sulla integrazione (in opposto alla specializzazione) tramite l'indipendenza delle definizioni date rispetto alle classi di guasti.

Questo documento puo' essere visto come un consenso minimo nella comunita' allo scopo di facilitare interazioni fruttuose; in aggiunta questo documento si spera sia adatto a) per essere usato da altri enti (incluse le organizzazioni per la standardizzazione), e b) per scopi educativi. In questo rispetto, lo sforzo terminologico qui presentato non e' fine a se stesso: le parole sono interessanti solamente in quanto catalogano univocamente dei concetti, e permettono di condividere idee e punti di vista. Questo documento non ha alcuna pretesa di essere uno stato dell'arte o una "Tavola dei Comandamenti": i concetti presentati devono evolvere con la tecnologia, e con il nostro progresso nella comprensione e nella padronanza della specifica, progetto e valutazione dei sistemi di elaborazione a funzionamento garantito.

Cio' che e' esposto in questo documento non sarebbe esistito senza le molte discussioni con molti colleghi, in particolare con i membri del Gruppo di Lavoro 10.4 dell'IFIP.

1 - DEFINIZIONI BASE

La **Garanzia di funzionamento (dependability)** e' definita come l'attendibilita' di un sistema di elaborazione tale che affidamento possa essere riposto in modo giustificato sul servizio che esso fornisce. Il **servizio (service)** fornito da un sistema e' il suo comportamento *cosi' come viene percepito* dal(i) suo(i) utente(i); un **utente (user)** e' un altro sistema (umano o fisico) che *interagisce* con il precedente.

A seconda della(e) applicazione(i) intesa(e) per il sistema, enfasi diversa puo' essere posta su aspetti differenti della garanzia di funzionamento, cioe' la garanzia di funzionamento puo' essere vista secondo *proprieta'* differenti, ma complementari, che permettono la definizione degli *attributi* della garanzia di funzionamento:

- rispetto all *prontezza per l'uso*, a funzionamento garantito significa **disponibile (available)**;
- rispetto alla *continuita' di servizio*, a funzionamento garantito significa **affidabile (reliable)**;
- rispetto all'*evitare conseguenze catastrofiche sull'ambiente,* a funzionamento garantito significa **a funzionamento sicuro (safe)**;
- rispetto alla *prevenzione di accessi e/o gestione dell'informazione non autorizzati,* a funzionamento garantito significa **sicuro (secure)**.

Un **fallimento (failure)** del sistema si verifica quando il servizio fornito non aderisce piu' alla **specifica**, essendo questa una descrizione concordata della funzione e/o servizio atteso del sistema. Un **errore (error)** e' quella parte dello stato del sistema che e' responsabile di portare ad un fallimento successivo: un errore che influenza il servizio e' una indicazione che un fallimento si verifica o si e' verificato. La causa aggiudicata o ipotizzata di un errore e' un **guasto (fault)**.

Lo sviluppo di un sistema di elaborazione a funzionamento garantito richiede l'utilizzazione *combinata* di un insieme di metodi che possono essere classificati in:

- **prevenzione del guasto (fault prevention)**: come prevenire il verificarsi o l'introduzione del guasto;
- **tolleranza al guasto (fault tolerance)**: come fornire un servizio conforme alla specifica nonostante i guasti;
- **eliminazione del guasto (fault removal)**: come ridurre la presenza (numero, gravita') dei guasti;
- **previsione del guasto (fault forecasting)**: come valutare il numero presente, l'incidenza futura, e le conseguenze dei guasti.

La prevenzione del guasto e la tolleranza al guasto possono essere viste come costituire il **conseguimento (procurement)** della garanzia di funzionamento: come *dotare* il sistema della capacita' di fornire un servizio in accordo con la specifica; la eliminazione del guasto e la previsione del guasto possono essere viste come costituire la **validazione (validation)** della garanzia di funzionamento: come *raggiungere la fiducia* nella capacita' del sistema di fornire un servizio in accordo con la specifica.

La fiducia nel servizio del sistema, e la giustificazione di tale fiducia, sono basate sulla *valutazione* del sistema, condotta principalmente rispetto agli *attributi* della garanzia di funzionamento, che sono pertinenti al servizio del sistema..

Le nozioni introdotte fino ad ora possono essere raggruppate in tre classi (figura 1):

- gli **impedimenti (impairments)** alla garanzia di funzionamento: guasti, errori, fallimenti; essi sono non desiderati - ma in principio non inattesi - circostanze che causano o che derivano dalla non garanzia di funzionamento (la cui definizione e' derivata molto semplicemente dalla definizione di garanzia di funzionamento: affidamento non puo', o non potra' piu' a lungo, essere posto sul servizio);
- i **mezzi (means)** per la garanzia di funzionamento: la prevenzione del guasto, la tolleranza al guasto, la eliminazione del guasto, la previsione del guasto; questi sono i metodi e le tecniche che permettono a) di dotare della capacita' di fornire un servizio su cui affidamento possa essere posto, e b) di raggiungere la fiducia in questa capacita'.
- gli **attributi (attributes)** della garanzia di funzionamento: disponibilita', affidabilita', sicurezza di funzionamento, sicurezza-confidenzialita'; questi mettono in grado a) di esprimere quali

proprieta' sono attese dal sistema, e b) di valutare la qualita' del sistema risultante dagli impedimenti e dai mezzi che li si oppongono.

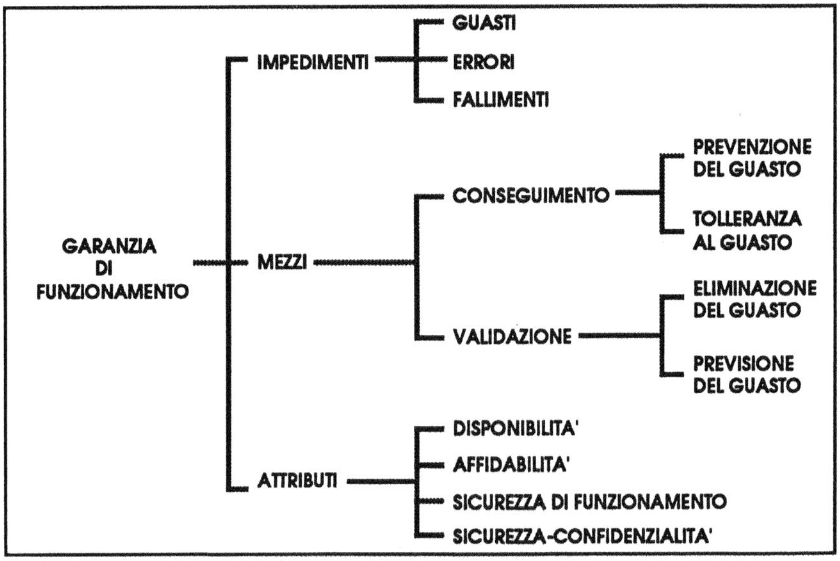

Figura 1 - L'albero della garanzia di funzionamento

2 - L'INTRODUZIONE DELLA GARANZIA DI FUNZIONAMENTO COME CONCETTO GENERICO

Una tendenza naturale di ogni disciplina emergente scientifica o tecnica e' inizialmente di restringere il suo campo di investigazione per poter fare - rapido - progresso nella soluzione dei problemi associati. Si giunge poi ad un punto in cui le sue interazioni con altre discipline non possono essere piu' ignorate. Una grossa tentazione e' allora quella di dichiarare le altre discipline "casi particolari" di quella considerata. Questo in genere da' luogo a molte dispute, spesso condotte dai sostenitori di ciascuna disciplina nel proprio gergo. Questo e' cio' che e' successo per l'affidabilita', la disponibilita', la sicurezza di funzionamento e la sicurezza-confidenzialita' dei sistemi di elaborazione. Inizialmente, il principale interesse era di far lavorare i sistemi di elaborazione: *l'affidabilita'*. Nel momento in cui i sistemi di elaborazione sono divenuti affidabili, i loro servizi sono stati usati e richiesti su base regolare, e cosi la *disponibilita'* e' divenuta essenziale. L'uso dei sistemi di elaborazione in applicazioni critiche ha introdotto l'interesse per la *sicurezza di funzionamento*. Il sistema meno pericoloso e' spesso quello che non fa niente, che non e' molto utile; pertanto, le persone interessate alla sicurezza di funzionamento tendono a considerare l'affidabilita' come sottoinsieme della sicurezza di funzionamento. L'avvento dei sistemi distribuiti ha esasperato gli aspetti di *sicurezza-confidenzialita'*. Nuovamente, un sistema sicuro deve poter adempiere delle funzioni; in aggiunta, le violazioni di sicurezza-confidenzialita' possono essere catastrofiche; pertanto, le persone interessate alla sicurezza-confidenzialita' tendono a considerare la sicurezza di funzionamento e l'affidabilita' come sottoinsiemi della sicurezza-confidenzialita'.

Comunque, le relazioni fra affidabilita', sicurezza di funzionamento e sicurezza-confidenzialita' sono piu' complesse di una semplice dipendenza. Consideriamo l'esempio delle cosiddette "bombe a tempo", cioe' guasti introdotti deliberatamente in un sistema di elaborazione per poter provocare, in un momento scelto dal "terrorista" - e sotto il suo controllo - un fallimento del sistema, con conseguenze preferibilmente percepite dall'utente come non catastrofiche (fintanto che egli non venga a conoscenza delle cause del fallimento). Questo esempio chiaramente coinvolge l'affidabilita', la sicurezza di funzionamento e la sicurezza-confidenzialita', in una maniera intricata e variabile dipendendo dai punti di vista considerati[1] . Cio' che e' certo e' che l'utente non puo', o non dovrebbe, fare affidamento sul servizio fornito da tale sistema, che non e' a funzionamento garantito nel vero senso del termine.

La discussione precedente mostra chiaramente che *non* e' intenzione di questo documento contribuire alla disputa se l'affidabilita' e' un concetto piu' ampio della sicurezza di funzionamento o viceversa, e similmente quando si consideri la sicurezza-confidenzialita'. Cio' che e' essenziale nella relazione fra affidabilita', sicurezza di funzionamento, sicurezza-confidenzialita' e garanzia di funzionamento e' che le prime tre sono punti di vista dell'ultima: pertanto, si spera che un mutuo scambio di esperienze venga favorito. Questa e' la principale ragione per l'aggiunta di un altro termine ad una lista gia' lunga - affidabilita', disponibilita', sicurezza di funzionamento, sicurezza-confidenzialita', etc. Un'altra ragione per l'introduzione di garanzia di funzionamento come concetto generico e' la volonta' di liberare il termine affidabilita' dal suo significato globale, etimologico - capacita' di potersi affidare - per concentrarsi sulla sua relazione al concetto di continuita' di servizio, accettata largamente (e recentemente), rispetto ad entrambi le sue

---

[1] Si noti anche che gli eventi riportati nella sezione su "Rischi al Pubblico nei Sistemi di Elaborazione" dell'ACM Software Engineering Notes riguardano l'affidabilita', la sicurezza di funzionamento e la sicurezza.

interpretazioni generali - di sistema affidabile - e di definizione probabilistica - affidabilita' di un sistema[2].

Sebbene "garanzia di funzionamento" sia sinonimo di affidabilita', esso ha una connotazione di dipendenza. Questo puo' essere percepito come connotazione negativa a prima vista, in confronto alla nozione positiva di fiducia espressa da affidabilita', ma mette in luce la dipendenza sempre crescente della nostra societa' da sistemi sofisticati in generale e particolarmente da sistemi di elaborazione. Continuando con considerazioni etimologiche, l'italiano affidabilita' e' tradotto dall'inglese "reliability", che deriva da "to rely" che viene dal francese "relier", derivato dal latino "religare", legare dietro: re - dietro e ligare, legare. La parola francese per affidabilita', "fiabilité", puo' essere ricondotta alla parola, del 12 secolo, "fiableté" il cui significato era " qualita' di essere attendibile"; l'italiano "affidabilita'" deriva dal latino "fidare", popolare di "fidere", che significa "avere fiducia, fidarsi". Alla luce di queste considerazioni etimologiche, ci si puo' solo rammaricare che la definizione di affidabilita' correntemente usata in molti campi dell'ingegneria[3] ha sostituito la nozione di "capacita'" alla nozione di "fiducia, attendibilita'", per (almeno) le due ragioni seguenti:

a) dal punto di vista dell'utente del sistema, cio' che e' di interesse in realta' non e' tanto la capacita' di fornire funzionalita', quanto il servizio che viene effettivamente fornito all'utente;
b) dal punto di vista del produttore del sistema che vuole ammettere la possibile esistenza di guasti nel suo progetto, l'interpretazione del termine "capacita'" deve essere ambigua, nonostante il fatto che essa sia stata adottata in glossari di ingegneria del software (vedi ad es. [IEE 82]).

---

2 E' interessante notare che:
a) la maggior parte dei libri che hanno il termine "affidabilita'" nel titolo trattano di come valutare, misurare, predire l'affidabilita' di sistemi, e non come costruire sistemi affidabili;
b) vedere la garanzia di funzionamento come un concetto piu' generale di affidabilita', disponibilita' etc. e incorporare questi ultimi termini, e' gia' stato tentato nel passato (vedi ad es. [Hos 60]); comunque quel tentativo era meno generale del nostro, poiche' lo scopo era di definire una misura che incorporasse la disponibilita' e la affidabilita', senza preoccuparsi della sicurezza.

3 Per esempio, "Reliability: The ability of an item to perform a required function under given conditions for a given time interval" [IEC 85].

3 - FUNZIONE, COMPORTAMENTO, STRUTTURA E SPECIFICA DEL SISTEMA

Fino ad ora, un **sistema (system)** e' stato - implicitamente - considerato nel suo complesso, mettendo in evidenza il suo comportamento percepito esternamente. Una definizione in accordo con questa visione "a scatola nera" e': una entita' che ha interagito o interferito, interagente o interferente, o in grado di interagire o interferire con altre entita', cioe' con altri sistemi. Questi altri sistemi sono stati, sono o saranno l'**ambiente (environment)** del sistema considerato[4].Un **utente** del sistema e' quella parte dell'ambiente che *interagisce* con il sistema considerato: l'utente fornisce ingressi a e/o riceve uscite dal sistema, avendo come caratteristica distintiva l'*uso del servizio* fornito dal sistema.

La **funzione (function)** di un sistema e' cio' che il sistema *e'inteso fare* [Kui 85]. Il **comportamento (behavior)** di un sistema e' cio' che il sistema *fa*. La **struttura (structure)** del sistema e' cio' che *gli fa fare cio' che fa* [Zie 76]. Adottando lo spirito di [And 81], un sistema, da un punto di vista strutturale ("a scatola aperta"), e' un insieme di componenti collegati insieme per poter interagire; un **componente (component)** e' un altro sistema, etc. La ricorsione si ferma quando un sistema viene considerato come **atomico (atomic)**: qualsiasi altra struttura interna non puo' essere distinta, o non e' di interesse e puo' essere ignorata. Il termine "componente" deve essere inteso in senso lato: livelli di un sistema come pure i componenti intralivello; in aggiunta, un componente essendo esso stesso un sistema, incorpora le interrelazioni dei componenti di cui e' composto. Una definizione piu' classica di struttura di un sistema e' cio' che il sistema *e'*. Tale definizione si adatta perfettamente quando si rappresenti un sistema senza tenere in conto esplicitamente alcun impedimento alla garanzia di funzionamento, e pertanto nel caso in cui la struttura e' considerata *fissa*. Noi non vogliamo limitarci a sistemi la cui struttura e' fissa. In particolare, dobbiamo permettere variazioni strutturali causate da, o risultanti da, impedimenti alla garanzia di

---

[4] a) Dare definizioni ricorsive non e' per gusto della ricorsione. Lo scopo e' di mettere in rilievo la relativita' rispetto al punto di vista adottato. Cosi' e' per la nozione di sistema: i confini di un dato sistema possono variare a seconda che sia visto dal(i) suo(i) progettista(i), dal(i) suo(i) utente(i), dal personale di manutenzione, etc.

b) Le forme passate, presenti e future sono usate per evidenziare che l'ambiente di un sistema puo' variare col tempo, specialmente rispetto alle fasi del suo ciclo di vita. Ad esempio, la nozione di "ambiente di programmazione" rientra nella definizione data, cosi' come l'ambiente fisico in cui un sistema deve operare.

funzionamento. Appare allora che una struttura possa avere stati[5]. Una definizione per la nozione di **stato (state)**: una condizione di essere rispetto ad un insieme di circostanze, *o di comportamento o di struttura.*[6].

Dalla sua definizione (comportamento percepito dall'utente), il servizio fornito da un sistema e' chiaramente una *astrazione* del comportamento del sistema. Si noti che questa astrazione dipende fortemente dall'applicazione che il sistema di elaborazione supporta. Un esempio di questa dipendenza e' il ruolo importante giocato in questa astrazione dal tempo: le granularita' temporali del sistema e del(i) suo(i) utente(i) sono in genere differenti, e la differenza varia da una applicazione all'altra. In aggiunta, il servizio e' naturalmente non ristretto solo alle uscite, ma comprende tutte le interazioni che sono di interesse per l'utente; ad esempio, la scansione dei sensori e' parte del servizio atteso da parte di un sistema di monitoraggio.

Fino ad ora abbiamo usato il singolare per funzione e servizio. Un sistema in generale esegue piu' di una funzione, e fornisce piu' di un servizio . La funzione ed il servizio possono essere visti come composti da singole funzioni e singoli servizi. Per semplicita', useremo il plurale - funzioni e servizi - quando si vogliano distinguere diverse singole funzioni e servizi.

Di particolare interesse per la garanzia di funzionamento sono le proprieta' temporali. Una **funzione o servizio in tempo reale (real-time function or service)** e' una funzione o servizio che deve essere fornito entro intervalli di tempo *dettati dall'ambiente*, ed un **sistema in tempo reale (real-time system)** e' un sistema che esegue almeno una funzione in tempo reale o fornisce almeno un servizio in tempo reale [PDC 90].

La **specifica (specification)** di un sistema descrive cio' che si aspetta che un sistema faccia in termini di a) sua funzione attesa e/o servizio atteso e di b) condizioni in cui - o sotto cui - essi devono essere eseguiti: ambiente, tempo di esposizione, prestazioni, osservabilita', etc. La funzione e/o il servizio sono di norma prima specificati in termini di cio' che *dovrebbe* essere eseguito o fornito per quel che riguarda gli scopi primari del sistema. Quando si considerino sistemi con caratteristiche di sicurezza di funzionamento o

---

5 a) Si potrebbe anche dire che una "struttura" ha anche um "comportamento", in particolare rispetto ai impedimenti della garanzia di funzionamento, anche se le velocita' di evoluzione considerate rispetto a) alla richiesta dell'utente da un lato, e b) agli impedimenti dall'altro, sono - sperabilmente - differenti.
b) la definizione data permette di includere altri tipi di sistemi con struttura variabile, ad esempio sistemi adattivi - in particolare quelli basati su conoscenza.

6 Questa definizione ha lo scopo di mettere in rilievo la relativita' della nozione di stato, che dipende direttamente dai fenomeni e dalle circostanze considerate; ad esempio stato rispetto all'attivita' di calcolo, stato rispetto al verificarsi di fallimenti.

sicurezza-confidenzialita', questa specifica e' in generale completata con cio' che *non dovrebbe* accadere (ad esempio gli stati di rischio da cui puo' derivare una catastrofe, o la divulgazione di informazioni riservate). Tale specifica puo' a sua volta portare a specificare - addizionali - funzioni o servizi che il sistema *dovrebbe* eseguire o fornire per ridurre la probabilita' di cio' che non dovrebbe accadere (ad esempio controllare i diritti di accesso di un utente ed autenticarlo).

In aggiunta, queste varie specifiche possono essere:

a) espresse in accordo a vari gradi di dettaglio: specifica dei requisiti, specifica di progetto, specifica di realizzazione, etc.
b) decomposte in accordo all'assenza o alla presenza di malfunzionamenti di componenti; il primo caso si riferisce a cio' che e' normalmente definito il modo di operazione *nominale*, e il secondo caso puo' essere in relazione al modo di operazione *degradato* se le risorse rimanenti non sono piu' a lungo sufficienti a fornire il servizio nominale.

Come conseguenza, non c'e' normalmente una singola specifica, ma diverse, e, chiaramente, un sistema puo' fallire rispetto ad alcune di queste specifiche, e ancora essere aderente alle altre.

E' essenziale che la specifica sia *concordata* fra due persone o enti legali: il fornitore del sistema (in senso lato: progettista, costruttore, venditore, etc.) e il suo utente umano[7]. L'accordo e' necessario perche' la specifica possa servire come base di giudizio se il servizio fornito e' accettabile o no, o, del pari, se un fallimento si e' verificato o no. Cio' che puo' essere giudicato come servizio accettabile rispetto ad una specifica ad un dato livello di dettaglio puo' non essere conforme con la specifica ad un livello meno dettagliato, a causa di errori verificatisi nel dettaglio della specifica, risultando in *errori di specifica*; gli errori di specifica possono a loro volta influenzare qualsiasi delle varie specifiche. Piu' in generale, non si puo' pretendere che una specifica sia immutabile una volta stabilita. Questo sarebbe semplicemente ignoranza dei fatti della vita, che implicano *cambiamenti*. I cambiamenti possono essere motivati dalla modifica dei requisiti del sistema: modifiche della funzione e/o del servizio, o correzione di qualche guasto[8]. Nuovamente, cio' che e' importante e' che la specifica sia concordata.

---

7 L'accordo puo' essere implicito, come quando si acquista un sistema che viene con la sua specifica e manuale d'utente, o quando si usino sistemi commerciali.

8 Ci troviamo di fronte ad un problema circolare: ci vuole un riferimento per giudicare se un servizio fornito e' accettabile o no, e questo riferimento puo' essere esso stesso guasto. E' stata dedicata molta attenzione al miglioramento delle specifiche, includendo proposte a questo scopo di modelli di cicli di vita [Boe 88].

Basandosi sul precedente punto di vista della struttura del sistema, le nozioni di funzione, di servizio e della loro specifica si applicano ugualmente in modo naturale ai componenti. Questo e' specialmente interessante nel processo di progetto, quando componenti commerciali, sia hardware che software sono usati: cio' che e' piu' di interesse per il progettista e' la funzione e/o il servizio che essi sono in grado di fornire, piuttosto che il loro comportamento (interno) dettagliato.

4 - GLI IMPEDIMENTI ALLA GARANZIA DI FUNZONAMIENTO

4.1 - Guasti (Faults)

I guasti e le loro cause sono molto diversi. Essi possono essere classificati secondo tre punti di vista principali che sono la loro natura, la loro origine e la loro persistenza.

La *natura* dei guasti porta a distinguere:

- **guasti accidentali (accidental faults)**, che si verificano o sono creati fortuitamente;
- **guasti intenzionali (intentional faults)**, che sono creati deliberatamente, presumibilmente in modo malevolo.

L'*origine* dei guasti puo' anche essa essere classificata secondo tre punti di vista:

- le *cause fenomenologiche*, che portano a distinguere [Avi 78]:
  - **guasti fisici (physical faults)**, che sono dovute a fenomeni fisici avversi,
  - **guasti causati dall'uomo (human-made faults)**, che sono dovuti alla imperfezione umana;
- i *confini del sistema*, che portano a distinguere:
  - **guasti interni (internal faults)**, che sono quelle parti dello stato del sistema che, quando richiamate dall'attivita' di elaborazione, produrranno un errore,
  - **guasti esterni (external faults)**, che derivano dall'interferenza dell'ambiente fisico nel sistema (perturbazioni elettromagnetiche, radiazioni, temperatura, vibrazioni, etc.), o dall'interazione con l'ambiente umano;

- la *fase di creazione* rispetto alla vita del sistema, che porta a distinguere:
  - **guasti di progetto (design faults)**, che derivano da imperfezioni che si verificano a) durante lo sviluppo del sistema (dalla specifica dei requisiti all'implementazione) o durante modifiche successive, o b) nella fase di stabilire le procedure di operazione o manutenzione del sistema;
  - **guasti operativi (operational faults)**, che si verificano durante l'uso del sistema.

Puo' essere fatta una distinzione anche rispetto alla *persistenza* temporale dei guasti, che porta a :

- **guasti permanenti (permanent faults)**, la cui *presenza* non e' in relazione a condizioni temporalmente puntuali, siano esse interne (attivita' di elaborazione) o esterne (ambiente),
- **guasti temporanei (temporary faults)**, la cui presenza e' in relazione a tali condizioni, e sono pertanto presenti per un periodo limitato di tempo.

Gli argomenti che interessano la sicurezza-confidenzialita' sono dominati da - ma non limitati a - guasti intenzionali, che sono chiaramente guasti *causati dall'uomo*. I guasti intenzionali possono essere sia interni che esterni; esempi tipici sono:

- per quel che riguarda i guasti interni, l'inserimento di **logica maliziosa (malicious logic)** (ad esempio i cosiddetti "cavalli di Troia"), che sono un guasto *di progetto* intenzionale;
- per quel che riguarda i guasti esterni, una **intrusione (intrusion)** che e' un guasto *operativo esterno*.

Perche' possano "avere successo", i guasti intenzionali possono avvantaggiarsi dei guasti accidentali , ad esempio una intrusione che sfrutta una breccia nella sicurezza-confidenzialita' causata da un guasto accidentale di progetto; ci sono somiglianze interessanti ed ovvie fra questo esempio ed un guasto temporaneo accidentale esterno che "sfrutta" una mancanza di schermatura.

Si potrebbe discutere sul fatto che introdurre le cause fenomenologiche nel criterio di classificazione dei guasti possa portare ricorsivamente molto indietro, ad esempio perche' i programmatori sbagliano? perche' falliscono i circuiti integrati? La nozione di guasto e' *arbitraria*, ed e' di fatto una possibilita' per terminare la ricorsione. Da qui la definizione data: causa *aggiudicata o ipotizzata* di un errore. Questa causa puo' variare a seconda del punto di vista considerato: meccanismi di tolleranza al guasto, ingegneri di manutenzione, negozio di riparazione, colui che ha sviluppato il sistema, fisico dei semiconduttori, etc. Dal nostro punto di vista, la ricorsione termina alla *causa che si intende prevenire o tollerare*. Questo punto di vista e' consistente con la distinzione fra guasti fisici e causati dall'uomo: un sistema di elaborazione e' un oggetto fatto dall'uomo e come tale qualsiasi guasto in esso o che lo riguarda e' in ultima analisi causato dall'uomo poiche' rappresenta l'incapacita' umana a dominare tutti i fenomeni che governano il comportamento di un sistema. In senso assoluto, una distinzione fra guasti fisici e causati dall'uomo (in particolare guasti di progetto) puo' non essere considerata necessaria; comunque, questa distinzione e' importante quando si considerino i metodi e le tecniche (attuali) per conseguire e validare la garanzia di funzionamento. Se la ricorsione detta in precedenza non viene terminata, allora *un guasto non e' altro che la conseguenza di un fallimento di qualche altro sistema* (incluso colui che lo ha sviluppato) *che ha fornito o sta ora fornendo un servizio al sistema dato.*

Seguono alcuni esempi della precedente discussione:

- un guasto di progetto deriva da un fallimento del progettista;
- un guasto fisico interno e' dovuto al malfunzionamento di un componente hardware, che a sua volta e' una conseguenza di (un) errori(e) a livello elettrico o elettronico (la comunita' di "affidabilita' in fisica" raramente caratterizza i guasti come "improvvisi ed impredicibili"), a loro volta originati da disordini fisico-chimici, a loro volta originati dalla produzione dell'hardware, o da - i limiti della - nostra conoscenza della fisica dei semiconduttori; i guasti interni fisici possono essere considerati come *ricorrenti (recurrent)* fintanto che la loro riparazione consiste nel rimpiazzare il componente fallito con uno identico, non fallito: il nuovo componente puo' fallire nuovamente ed in modo simile nel futuro a meno che le cause del fallimento siano tracciate nei processi di progetto e produzione e rimosse, portando cosi' ad un componente modificato con una probabilita' di guasto inferiore;
- un guasto esterno fisico o causato dall'uomo e' di fatto un guasto di progetto: l'incapacita' di prevedere tutte le situazioni che il sistema incontrera' durante la sua vita operativa, o il rifiuto di considerare alcune di esse (ad esempio per ragioni economiche); ad esempio:
  - nel caso di una perturbazione elettromagnetica: e' un guasto esterno o un guasto di progetto, cioe' la mancanza di adeguata schermatura?

- nel caso di un fallimento causato da un operatore che batte su una tastiera un singolo carattere sbagliato: e' un guasto di interazione o un guasto di progetto, cioe' mancanza della conferma richiesta dal sistema [Nor 83]?

Il punto di vista della persistenza temporale merita i seguenti commenti:

1) I guasti esterni temporanei che originano dall'ambiente fisico sono spesso chiamati **guasti transitori (transient faults)**.
2) I guasti interni temporanei sono spesso chiamati **guasti intermittenti (intermittent faults)**; tali guasti derivano dalla presenza di combinazioni di condizioni che si verificano raramente; esempi sono a) guasti "sensibili allo schema" in memorie a semiconduttore, cambiamenti nei parametri di un componente hardware (effetto della variazione di temperatura, ritardo dovuto a capacita' parassite, etc.), o b) situazioni - che influenzano l'hardware o il software - che si verificano quando il carico del sistema va oltre un certo livello, del tipo di temporizzazioni e sincronizzazione critici. Il termine "guasto", infatti, in tali casi e' di fatto una astrazione per *condizioni di guasto*. La nozione di guasto intermittente e' in senso assoluto arbitraria: tali guasti non sono altro che guasti (permanenti) la cui condizione di attivazione non puo' essere riprodotta o che si verifica abbastanza raramente; comunque, come gia' rilevato a proposito della distinzione fra guasti fisici e di progetto, considerarli e' di utilita'.

Dalla precedente discussione, appare che *qualunque guasto puo' essere considerato un guasto di progetto permanente*. Questo e' vero in assoluto, ma non e' di molto aiuto per chi deve sviluppare e valutare un sistema.

La figura 2 riassume le varie classi di guasto che abbiamo trattato, rispetto ai vari punti di vista che abbiamo considerato.

Se tutte le combinazioni di classi di guasto secondo i 5 punti di vista di figura 2 fossero possibili, ci sarebbero 32 classi di guasto differenti. Di fatto, il numero di combinazioni possibili e' piu' limitato: 11 combinazioni sono indicate dalle righe di figura 3, che da' anche il nome usuale di queste combinazioni, *non le loro definizioni*. Questi nomi sono comunemente usati per esprimere in maniera sintetica i risultati della combinazione di punti di vista differenti.

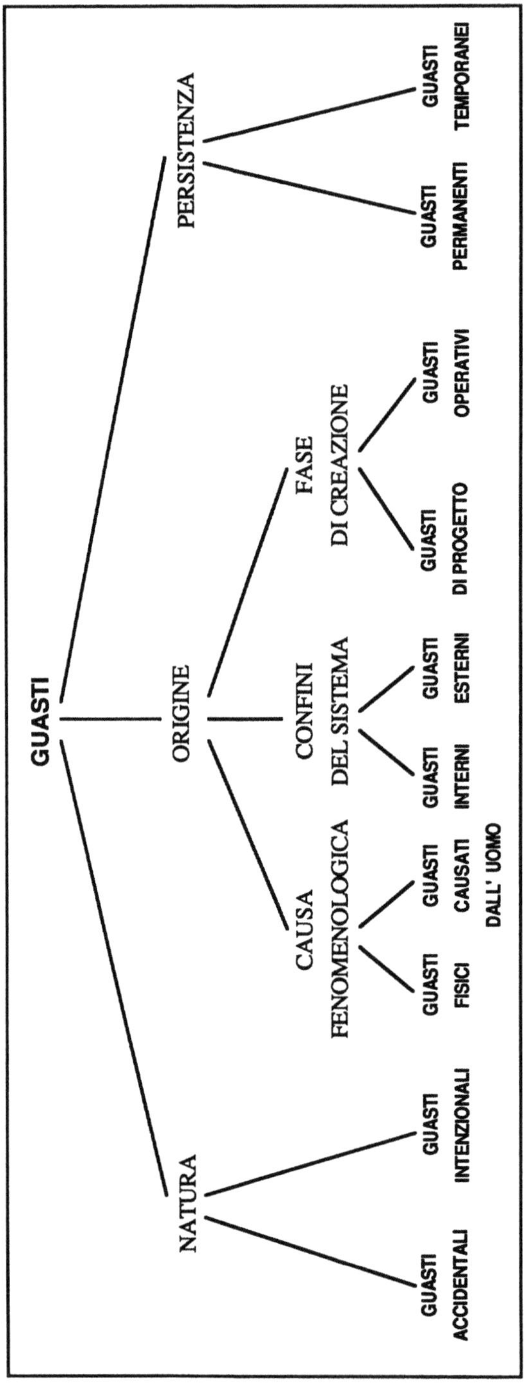

Figura 2 - Le classi di guasto secondo vari punti di vista

Natura		Origine						Persistenza		Nome Usuale
		Causa Fenomenologica		Confini del Sistema		Fase di Creazione				
Guasti Accidentali	Guasti Intenzionali	Guasti Fisici	Guasti causati dall' Uomo	Guasti Interni	Guasti Esterni	Guasti di Progetto	Guasti Operativi	Guasti Permanenti	Guasti Temporanei	
✓		✓		✓			✓	✓		*Guasti Fisici*
✓		✓			✓		✓	✓		
✓		✓			✓		✓		✓	*Guasti Transitori*
✓		✓		✓			✓		✓	*Guasti Intermittenti*
✓			✓	✓		✓			✓	
✓			✓	✓		✓		✓		*Guasti di Progetto*
✓			✓		✓		✓		✓	*Guasti di Interazione*
	✓		✓	✓		✓		✓		*Logica Maliziosa*
	✓		✓	✓		✓			✓	
	✓		✓		✓		✓	✓		*Intrusioni*
	✓		✓		✓		✓		✓	

Figura 3 - Le classi di guasti che risultano da combinazioni secondo i vari punti di vista di figura 2

4.2 - Errori (Errors)

Un errore e' stato definito come essere responsabile di condurre ad un fallimento successivo. Se un errore portera' effettivamente o no ad un fallimento dipende da tre fattori principali:

1) La composizione del sistema, e specialmente la natura della ridondanza esistente:
   - ridondanza *intenzionale* (introdotta per fornire tolleranza al guasto) che e' esplicitamente intesa per prevenire che un errore conduca ad un fallimento,
   - ridondanza *non intenzionale* (e' difficile in pratica, se non impossibile, costruire un sistema senza alcuna forma di ridondanza[9]) che puo' avere lo stesso risultato - non atteso - della ridondanza intenzionale.
2) L'attivita' del sistema: un errore puo' essere compensato prima di provocare danno.
3) La definizione di un fallimento dal punto di vista dell'utente: cio' che e' un fallimento per un dato utente puo' essere una sopportabile noia per un altro. Esempi sono a) tenere in conto la granularita' temporale dell'utente: un errore che "transita" nell'interfaccia sistema-utente(i) puo' essere visto o no come un fallimento a seconda della granularita' temporale dell'utente, b) la nozione di "frequenza accettabile di errore" - implicitamente prima di considerare che un errore si e' verificato - nella trasmissione dati. Questa discussione spiega perche' e' spesso desiderabile menzionare esplicitamente nella specifica condizioni quali il tempo massimo di interruzione (relativo alla granularita' temporale dell'utente).

4.3 Fallimenti o (Malfunzionamenti) (Failures)

Basandosi sulla definizione data di struttura di sistema, la discussione se "fallimento" si applichi a un sistema o ad un componente e' semplicemente irrilevante, poiche' un componente e' esso stesso un sistema. Quando si tratta con sistemi atomici, la nozione di fallimento "elementare" viene naturale.

Un sistema puo' non fallire, e generalmente non fallisce, sempre nello stesso modo. I modi in cui un sistema puo' fallire sono i suoi *modi di fallimento*, che

---

9 Un problema classico nel collaudo dell'hardware e' la rimozione di tali "false ridondanze", il cui effetto puo' essere il mascheramento dei guasti, e come tale rendere il compito di generare configurazioni di test piu' complicato.

possono essere caratterizzati secondo tre punti di vista: dominio, percezione da parte dell'utente del sistema, e conseguenze sull'ambiente.

Il punto di vista del *dominio di fallimento* porta a distinguere:

- **fallimenti nel valore (value failures)**: il valore del servizio fornito non e' conforme con la specifica;
- **fallimenti nel tempo (timing failures)**: la temporizzazione della fornitura del servizio non e' conforme con la specifica.

Tali definizioni generali (non conformita' con la specifica) si applicano a **fallimenti arbitrari (arbitrary failures)**. Si possono distinguere modi piu' fini di fallimento. Ad esempio, la nozione di fallimento nel tempo puo' essere affinata in fallimento nel tempo *per anticipo (early timing failures)* o fallimenti nel tempo *per ritardo (late timing failures)*, a seconda se il servizio e' fornito troppo presto o troppo tardi. Una classe di fallimenti che si riferiscono sia al dominio del valore che del tempo sono i **fallimenti con blocco (stopping failures)**: l'attivita' del sistema non e' piu' percepibile dagli utenti, ed un servizio a valore costante viene fornito; il valore costante fornito puo' variare secondo l'applicazione, ad esempio l'ultimo valore corretto, un valore predeterminato, etc. un caso particolare di fallimenti con blocco e' quello dei **fallimenti per omissione (omission failures)** [Cri 85b, Ezh 86]: nessun servizio e' fornito. Un tale fallimento puo' essere considerato come caso limite comune sia per fallimenti nel valore (valore nullo) che per fallimenti nel tempo (fallimento per ritardo infinito); un fallimento per omissione *persistente* e' un **fallimento per omissione persistente (crash failure)**. Un sistema i cui fallimenti possono essere solamente - o piu' in generale entro un certo limite accettabile - fallimenti con blocco, e' un **sistema a fallimento con blocco (fail-stop system)**, ed un sistema i cui fallimenti possono essere solamente - o entro limiti accettabili - fallimenti per omissione persistente, e' un **sistema a fallimento tacito (fail-silent system)** [Pow 88][10] .

Quando un sistema ha diversi utenti, il punto di vista della *percezione del fallimento* porta a distinguere:

- **fallimenti consistenti (consistent failures)**: tutti gli utenti del sistema hanno la stessa percezione dei fallimenti;
- **fallimenti inconsistenti (inconsistent failures)**: gli utenti del sistema possono avere percezioni differenti di un dato fallimento; i

---

10 La nozione di *processore* a fallimento con blocco, definita in [Sch 83] nel contesto dei sistemi distribuiti puo' essere considerata un esempio di sistema a fallimento con blocco.

fallimenti inconsistenti sono usualmente chiamati, seguendo [Lam 82], *fallimenti bizantini (Byzantine failures).*

La **gravita' (severity)** del fallimento risulta dalla graduazione delle *conseguenze dei fallimenti* sull'ambiente del sistema. Essa permette pertanto un ordinamento dei modi di fallimento. Un caso particolare di grande interesse e' quello di sistemi i cui modi di fallimento possono essere raggruppati in due classi di gravita' considerevolmente differente:

- **fallimenti benigni (benign failures)**, per cui le conseguenze sono dello stesso ordine di grandezza (in genere in termini di costo) del beneficio prodotto dal servizio fornito in assenza di fallimento;
- **fallimenti catastrofici (catastrophic failures)**, per cui le conseguenze sono incommensurabilmente piu' grandi del beneficio prodotto dal servizio fornito in assenza di fallimento.

Un sistema i cui fallimenti possono essere solamente - o piu' in generale entro limiti accettabili - fallimenti benigni e' un **sistema a fallimento non pericoloso (fail-safe system)**. La nozione di gravita' di fallimento permette di definire il concetto di criticita': la **criticita' (criticality)** di un sistema e' la gravita' piu' elevata dei suoi (possibili) modi di fallimento[11]. La relazione fra modi di fallimento e gravita' di fallimento dipende fortemente dall'applicazione. Comunque, esiste una ampia classe di applicazioni in cui la non operazione e' considerata una posizione naturalmente non pericolosa (ad esempio trasporto a terra, produzione di energia), da cui la diretta corrispondenza che spesso si fa fra fallimento con blocco e fallimento non pericoloso [Min 67, Nic 89].

4.4 - Patologia del guasto

I meccanismi di creazione e manifestazione di guasti, errori e fallimenti possono riassumersi come segue:

1) un guasto e' **attivo (active)** quando produce un errore. Un guasto attivo e' o a) un guasto interno che era in precedenza **inattivo (dormant)** e che e' stato attivato dal processo di elaborazione (includendo l'esistenza simultanea delle condizioni di guasto per un guasto intermittente), o b) un guasto esterno. La maggior parte dei

[11] Come esempio, i livelli di criticita' accettati in avionica sono definiti come segue [RTC 85]:
- critico: funzioni per cui l'occorrenza di un qualsiasi fallimento impedisce la continuazione non pericolosa del volo e dell'atterraggio dell'aereo;
- essenziale: funzioni per cui l'occorrenza di un qualsiasi fallimento riduce la capacita' dell'aereo o la possibilita' dell'equipaggio a fronteggiare condizioni operative avverse;
- non essenziale: funzioni per cui un fallimento non degrada significativamente la capacita dell'aereo o la capacita' di azione dell'equipaggio.

guasti interni puo' ciclare fra lo stato attivo e lo stato inattivo. I guasti fisici possono influenzare direttamente soltanto i componenti hardware, mentre i guasti causati dall'uomo possono influenzare qualsiasi componente.

2) Un errore puo' essere latente o rilevato. Un errore e' **latente (latent)** quando non e' stato riconosciuto come tale; un errore e' **rilevato (detected)** da un algoritmo o meccanismo di rilevazione. Un errore puo' scomparire prima di essere rilevato. Un errore puo' propagarsi, ed in generale si propaga; propagandosi, un errore crea altri - nuovi - errori.

3) Un fallimento si verifica quando un errore "passa attraverso" l'interfaccia sistema-utente ed influenza il servizio fornito dal sistema. Il fallimento di un componente da' luogo ad un guasto a) nel sistema che contiene quel componente, e b) dal punto di vista degli altri componenti con cui interagisce; i modi di fallimento del componente fallito divengono allora tipi di guasto per i componenti che interagiscono con esso.

Questi meccanismi permettono di completare la "catena fondamentale":

... ---> fallimento ---> guasto ---> errore ---> fallimento ---> guasto ---> ...

Alcuni esempi illustrativi della patologia del guasto:

- il risultato di un *errore* di un programmatore e' un *guasto (inattivo)* nel software scritto (istruzione(i) o dati guasti); all'attivazione (invocando il componente in cui risiede il guasto ed attivando l'istruzione guasta, o la sequenza di istruzioni o i dati con una opportuna configurazione di ingresso) il guasto diviene *attivo* e produce un errore; se e quando il dato erroneo influenza il servizio fornito (in valore e/o nella temporizzazione della fornitura), si verifica un *fallimento*;
- un corto circuito che si verifica in un circuito integrato e' un *fallimento* (rispetto alla specifica del servizio del circuito); la conseguenza (una connessione bloccata ad un valore booleano, la modifica della funzione del circuito, etc.) e' un *guasto* che rimarra' inattivo fintanto che non viene attivato, essendo identica al precedente esempio la continuazione del processo;
- una perturbazione elettromagnetica di sufficiente energia e' un *guasto*; questo guasto puo'
  - a) creare direttamente un *errore*, ad esempio per interferenza elettromagnetica con le cariche elettriche che circolano nei conduttori,

b) creare un altro guasto (interno); ad esempio, se la perturbazione agisce sull'ingresso di memoria in posizione di scrittura cambiando qualche cifra binaria, questi errori rimarranno come guasti in memoria; essi rimarranno inattivi fintanto che la particolare posizione di memoria viene letta; la sequenza errore-fallimento a partire dal guasto transitorio esterno al guasto interno esiste ancora, a livello elettronico;

- una interazione scorretta uomo macchina eseguita da un operatore durante l'operazione del sistema e' un *guasto* (dal punto di vista del sistema); il dato alterato risultante elaborato e' un *errore*; etc.
- un errore di chi scrive un manuale di manutenzione o operativo puo' dar luogo ad un guasto nel corrispondente manuale (direttive sbagliate) che rimangono inattive fintanto che le direttive non sono utilizzate per trattare una qualche situazione, etc.

Dagli esempi precedenti, si comprende facilmente che l'inattivita' del guasto puo' variare considerevolmente, a seconda del guasto, della utilizzazione del dato sistema, etc.

I guasti causati dall'uomo possono essere accidentali o intenzionali. L'esempio precedente relativo all'errore del programmatore e le sue conseguenze possono essere rifrasati come segue: una bomba logica viene creata da un programmatore malizioso; essa rimarra' inattiva fino a quando non sara' attivata (ad esempio a qualche momento predeterminato); essa produrra' un errore che puo' portare a supero di memoria o al rallentamento delle esecuzioni dei programmi; come conseguenza, il servizio fornito soffrira' di un cosiddetto rifiuto di servizio, un tipo particolare di fallimento.

Questi esempi sono stati mantenuti semplici di proposito. La vita reale e' di solito molto piu' complessa; quattro esempi:

a) un dato guasto in un dato componente puo' originare da diverse sorgenti possibili; ad esempio, un guasto permanente in un componente fisico - ad esempio bloccato alla tensione di massa - puo' originare da:
   - un fallimento fisico (ad esempio causato dal cambiamento di una soglia di tensione),
   - un errore causato da un guasto di progetto - ad esempio una microistruzione sbagliata - che si propaga "dall'alto al basso" attraverso i livelli e che causa un corto circuito fra due uscite di un circuito per un tempo abbastanza lungo da provocare un corto circuito che ha le stesse conseguenze del cambio di soglia;

b) un guasto di una data classe puo', per mezzo della propagazione di errore, creare un guasto di un'altra classe; ad esempio, il guasto che ha portato ad un errore durante l'esecuzione della microistruzione dell'esempio precedente potrebbe essere stato un guasto transitorio;
c) alcuni punti di vista possono diventare - almeno temporaneamente - di minore importanza durante il processo di propagazione; ad esempio, quando si tratti con guasti esterni che producono errori di ingresso durante l'esecuzione di un componente software (pertanto invocandolo nel cosiddetto dominio eccezionale di ingresso [Cri 80]), il fatto che il guasto sia fisico o causato dall'uomo puo' non essere rilevante per il comportamento di fallimento del dato componente;
d) un fallimento spesso risulta dall'azione combinata di diversi guasti; questo e' vero specialmente quando si considerino argomenti di sicurezza-confidenzialita': un passaggio-trappola (cioe' qualche mezzo di superare il controllo dell'accesso) che sia inserito in un sistema di elaborazione, o accidentalmente o intenzionalmente, e' un guasto di progetto; questo guasto puo' rimanere inattivo fino a quando qualche persona maliziosa fa uso di esso per introdursi nel sistema; una connessione intrusiva e' un guasto di interazione intenzionale ; quando l'intruso e' connesso (mentre non dovrebbe), puo' deliberatamente creare un errore, ad esempio modificando qualche file (attacco all'integrita'); quando questo file e' usato da un utente autorizzato, verra' influenzato il servizio, e si verifichera' un fallimento.

Due commenti aggiuntivi, a proposito delle parole, o termini, "guasto", "errore" e "fallimento":

a) il loro uso esclusivo in questo lavoro non preclude l'uso, in situazioni particolari, di parole che designano, in modo compatto e non ambiguamente, una classe particolare di impedimenti; questo e' applicabile in particolare ai guasti (ad esempio bug[12], difetti, deficienze) ed ai fallimenti (ad esempio rotture, malfunzionamenti, rifiuti di servizio);
b) L'assegnazione fatta del termine particolare guasto, errore, fallimento, prende in considerazione semplicemente l'uso corrente: i) prevenzione, tolleranza e diagnosi del guasto, ii) rilevazione e correzione dell'errore, iii) frequenza di fallimento.

---

12 Includendo anche la specializzazione del termine "bug", come in [Gra 86], che distingue "Heisenbugs" (guasti software intermittenti, dal principio di indeterminazione di Heisenberg) e "Bohrbugs" (guasti software permanenti, "come l'atomo di Bohr, solido, facilmente rilevato con tecniche standard, e pertanto noioso (boring)")

Infine si deve metter in rilievo che le definizioni date in questo capitolo sono *sintattiche*; conseguentemente, sono stai enfatizzati i criteri relativi alle varie classificazioni, ed essi sono secondo il nostro punto di vista piu' importanti delle classi stesse.

5 - I Mezzi per la Garanzia di Funzionamento

5.1 - Dipendenze fra i mezzi per la garanzia di funzionamento

Tutti i "come" che appaiono nelle definizioni di base date nel capitolo 1 sono di fatto obiettivi che non possono essere raggiunti completamente, in quanto tutte le attivita' corrispondenti sono attivita' umane, e pertanto imperfette. Queste imperfezioni introducono delle *dipendenze* che spiegano perche' e' soltanto l'uso *combinato* dei metodi predetti - preferibilmente a tutti i passi dei processi di progetto ed implementazione - che possono portare ad un sistema caratterizzato da garanzia di funzionamento. Queste dipendenze possono essere schematizzate come segue: nonostante la prevenzione del guasto per mezzo di metodologie di progetto e regole di costruzione (imperfette perche' siano possibili), i guasti si verificano. Pertanto c'e' necessita' della eliminazione del guasto. Anche essa e' imperfetta, cosi' come sono imperfetti i componenti commerciali - hardware e software - del sistema, e pertanto e' importante la previsione del guasto. La nostra crescente dipendenza dai sistemi di elaborazione porta al requisito della tolleranza al guasto, che a sua volta si basa su regole di costruzione; pertanto da capo eliminazione del guasto, previsione del guasto, etc. Si noti che il processo e' anche piu' ricorsivo di quanto appaia da cio' che precede: gli attuali sistemi di elaborazione sono cosi' complessi che il loro progetto richiede strumenti computerizzati per poter essere economici. Anche questi strumenti devono essere a funzionamento garantito, e cosi' via.

Il precedente ragionamento illustra la stretta interazione fra la eliminazione del guasto e la previsione del guasto, e motiva il raccoglierli sotto il termine unico di *validazione*. Questo nonostante il fatto che la validazione e' spesso limitata alla eliminazione del guasto, ed associata con una delle principali attivita' coinvolte nella eliminazione del guasto, la verifica: ad esempio in "V and V" [Boe 79]; in tale caso la distinzione e' relativa alla differenza fra "costruire il sistema in modo corretto" (relativa alla verifica) e "costruire il corretto sistema" (relativa alla validazione)[13]. Cio' che si propone qui e' semplicemente una estensione di questo concetto: la risposta alla domanda "sto costruendo il sistema corretto?" (eliminazione del guasto) essendo complementata da "per quanto tempo sara' corretto?" (previsione del guasto)[14]. In aggiunta, la eliminazione del guasto e' usualmente strettamente associata con la prevenzione del guasto, formando insieme il **modo di evitare il guasto (fault avoidance)**, cioe' come *mirare* ad un sistema esente da guasto. Oltre a mettere in luce la necessita' di validare le procedure ed i meccanismi della tolleranza al guasto, considerare la eliminazione del guasto e la previsione del guasto come due costituenti la stessa attivita' - la validazione - e' di grande interesse in quanto permette una migliore comprensione della nozione di copertura, e pertanto di un importante problema introdotto dalla ricorsione precedente: *la validazione della validazione*, ovvero come raggiungere la confidenza nei metodi e negli strumenti usati per avere confidenza nel sistema. La **copertura (coverage)** si riferisce qui ad una misura della rappresentativita' delle situazioni a cui e' sottoposto il sistema durante la validazione, in confronto alle effettive situazioni in cui si trovera' durante la sua vita operativa[15]. Il concetto di copertura imperfetta rafforza la relazione fra eliminazione del guasto e previsione del guasto, in quanto si puo' considerare che la necessita' della previsione del guasto deriva dalla copertura imperfetta della eliminazione del guasto.

Nel seguito di questo capitolo, noi esaminiamo la tolleranza al guasto, la eliminazione del guasto e la previsione del guasto; la prevenzione del guasto non viene trattata poiche' si riferisce alla ingegnerizzazione "in generale" del sistema.

5.2 - Tolleranza al guasto

La tolleranza al guasto [Avi 67] e' ottenuta per mezzo del trattamento dell'errore e del trattamento del guasto [And 81]. Il **trattamento dell'errore**

---

13 E' degno di nota che questi assegnamenti sono talvolta invertiti, come ad esempio nel dominio dei protocolli di comunicazione (vedi ad esempio [Rud 85]).

14 Validazione deriva da "validita'", che incapsula due nozioni:
- validita' ad un dato momento, che e' relativa alla rimozione del guasto;
- validita' per una data durata, che e' relativa alla previsione del guasto.

15 La nozione di copertura qui definita e' molto generale; puo' essere resa piu' precisa indicando i suoi campi di applicazione, ad esempio:
- copertura di un test software rispetto al codice, grafo di controllo etc,;
- copertura di un test di un circuito integrato rispetto ad un modello di guasto;
- copertura della tolleranza al guasto rispetto ad una classe di guasti;
- copertura di una ipotesi di progetto rispetto alla realta'.

(error processing) ha lo scopo di rimuovere gli errori dallo stato della computazione, possibilmente prima del verificarsi di un fallimento; il **trattamento del guasto (fault treatment)** ha lo scopo di prevenire la attivazione - ulteriore - dei guasti.

Il *trattamento dell'errore* puo' essere eseguito in due modi:

- **recupero dall'errore (error recovery)**, con cui uno stato esente da errore viene sostituito allo stato erroneo; questa sostituzione puo' avvenire in due modi [And 81]:
  - **recupero indietro (backward recovery)**, in cui la trasformazione dello stato erroneo consiste nel riportare il sistema in uno stato gia' occupato precedentemente al verificarsi dell'errore; questo coinvolge la determinazione di **punti di recupero (recovery points)**, che sono punti sulla scala temporale relativa all'esecuzione di un processo per cui un precedente stato corrente possa essere successivamente ripristinato;
  - **recupero in avanti (forward recovery)**, in cui la trasformazione dello stato erroneo consiste nel trovare un nuovo stato, a partire dal quale il sistema possa operare (frequentemente in modo degradato);
- **compensazione dell'errore (error compensation)**, in cui lo stato erroneo contiene sufficiente informazione (ridondanza) per permettere la fornitura di un servizio esente da errore a partire dallo stato (interno) erroneo.

Quando si usa il recupero dall'errore, e' necessario che lo stato erroneo venga (rapidamente) identificato come tale prima di essere trasformato; questo e' lo scopo della **rilevazione dell'errore (error detection)**, da cui il termine usualmente usato di *rilevazione dell'errore e recupero*. L'associazione in un componente della sua capacita' funzionale insieme con i meccanismi di rilevazione dell'errore porta alla nozione di **componente che si autocontrolla (self-checking component)**, o in hardware [Car 68, Wak 78, Nic 89] o in software [Yau 75, Lap 90a]; uno dei piu' importanti benefici dell'approccio basato su componenti che si autocontrollano e' la possibilita' di dare una chiara definizione di *aree di confinamento dell'errore* [Sie 82]. Quando si esegue una compensazione dell'errore in un sistema realizzato con componenti che si autocontrollano suddivisi in classi che eseguono gli stessi compiti, allora la trasformazione dello stato non e' altro che la commutazione entro una classe da un componente fallito ad uno non fallito, da cui il corrispondente approccio alla tolleranza al guasto: *rilevazione* *dell'errore e compensazione*[16]. D'altra parte, la compensazione puo' essere applicata sistematicamente, anche in assenza di errori, ottenendo il **mascheramento del guasto (fault masking)** (ad esempio nella votazione con maggioranza). Comunque, questo puo' corrispondere allo stesso tempo ad una diminuzione di ridondanza non nota. Pertanto implementazioni pratiche del mascheramento coinvolgono in genere la rilevazione dell'errore, che puo' essere eseguita *dopo* la trasformazione dello stato.

Il recupero indietro dall'errore e quello in avanti non sono mutuamente esclusivi: puo' essere prima tentato un recupero indietro; se l'errore persiste, puo' essere tentato il recupero in avanti. Nel recupero in avanti, e' necessario *valutare il danno (assess the damage)* causato dall'errore rilevato, o dagli errori propagati prima della rilevazione; la valutazione del danno puo' - in linea di principio - essere ignorata nel caso di recupero indietro, purche' i meccanismi di trasformazione da uno stato erroneo ad uno stato esente da errore non siano stati influenzati [And 81].

Il tempo aggiuntivo necessario per il trattamento di errore dipende fortemente dal tipo di trattamento di errore adottato:

- nel recupero dall'errore, il tempo aggiuntivo e' maggiore quando si verifica un errore rispetto a prima; particolarmente, nel recupero indietro e' relativo allo stabilire i punti di recupero, per prepararsi al trattamento dell'errore;
- nella compensazione dell'errore, il tempo aggiuntivo di cui necessita la compensazione e' lo stesso, o quasi, sia se gli errori sono presenti sia che siano assenti[17].

In aggiunta, la durata della compensazione dell'errore e' molto piu' breve della durata del recupero da errore, a causa della maggiore quantita' di ridondanza (strutturale).

Questa osservazione

a) e' di importanza pratica in quanto condiziona spesso la scelta della strategia adottata di tolleranza al guasto rispetto alla granularita' temporale dell'utente;
b) ha introdotto una relazione fra il tempo aggiuntivo e la ridondanza strutturale; piu' in generale, un sistema ridondante ha sempre un

---

16 La rilevazione dell'errore e compensazione possono essere considerate come caso limite della rilevazione dell'errore e recupero, dove il recupero e' effettuato usando lo stato presente (erroneo) del sistema invece che sostituirgli uno stato esente da errore.

17 In entrambi i casi, il tempo per aggiornare lo stato del sistema si somma al tempo aggiuntivo.

comportamento ridondante, caratterizzato da un tempo aggiuntivo di operazione; il tempo aggiuntivo puo' essere tanto piccolo da non essere percepibile dall'utente, il che significa soltanto che il *servizio* non e' ridondante; un estremo opposto e la "ridondanza temporale" (*comportamento* ridondante ottenuto tramite ripetizione) che deve essere almeno inizializzata da una ridondanza strutturale, limitata ma pur sempre presente; con una certa approssimazione si puo' dire che maggiore e' la ridondanza strutturale, minore e' il tempo aggiuntivo di operazione necessario.

Il primo passo nel *trattamento del guasto* e' la **diagnosi del guasto (fault diagnosis)** che consiste nel determinare la(e) causa(e) dell'(degli) errore(i), sia in termini di locazione che di natura. Seguono poi le azioni tese ad ottenere lo scopo principale del trattamento del guasto: prevenire che il(i) guasto(i) sia nuovamente attivato, tendendo a renderlo passivo, e cioe' la **disattivazione del guasto (fault passivation)**. Cio' e' ottenuto rimuovendo il(i) componente(i) identificati come guasti da ulteriori esecuzioni. Se il sistema non e' piu' a lungo capace di fornire lo stesso servizio di prima, allora puo' aver luogo una *riconfigurazione (reconfiguration).*

Se si valuta che il trattamento dell'errore puo' rimuovere direttamente il guasto, o se la probabilita' di ripresentarsi e' abbastanza bassa, allora la disattivazione del guasto puo' non essere eseguita. Fintanto che la disattivazione del guasto non e' eseguita, il guasto viene considerato come **guasto debole (soft fault)**; eseguire la disattivazione del guasto implica che esso viene considerato come **forte (hard)**, o **solido (solid)**. A prima vista, le nozioni di guasto debole e forte possono apparire sinonimi delle nozioni precedentemente introdotte di guasto temporaneo e permanente. Invero, la tolleranza ai guasti temporanei non ha bisogno del trattamento del guasto, poiche' il recupero dall'errore dovrebbe, in questo caso, rimuovere direttamente gli effetti del guasto, che e' gia' sparito, purche' un guasto permanente non sia stato creato nel processo di propagazione. Infatti, le nozioni di guasti deboli e forti sono utili per i seguenti motivi:

- distinguere un guasto permanente da uno temporaneo e' un compito difficile e complesso, poiche' a) un guasto temporaneo sparisce dopo un certo tempo, usualmente prima che sia eseguita una diagnosi del guasto, e b) guasti appartenenti a classi differenti possono portare ad errori molto simili; cosi', la nozione di guasto debole o forte di fatto incorpora la soggettivita' associata a queste difficolta', includendo il fatto che un guasto possa essere dichiarato come debole quando la diagnosi del guasto non ha successo;

- la capacita' di queste nozioni di incorporare sottigliezze dei modi di azione di alcuni guasti transitori; ad esempio, si puo' dire che il guasto interno inattivo che risulta dall'azione di particelle alfa (causate dalla ionizzazione residua degli involucri circuitali), o di ioni pesanti nello spazio, su elementi di memoria (in senso lato, includendo anche flip-flops) e' un guasto *temporaneo* ?Tale guasto inattivo e' comunque un guasto *debole.*

Le definizioni precedenti si applicano sia ai guasti fisici che ai guasti di progetto: la(e) classe(i) di guasti che puo' essere tollerata effettivamente dipende(ono)dalle ipotesi di guasto che vengono considerate nel processo di progetto, e si basa pertanto sulla *indipendenza* delle ridondanze rispetto ai processi di creazione e di attivazione del guasto. Un esempio e' fornito dal considerare la tolleranza ai guasti fisici e la tolleranza ai guasti di progetto. Un metodo (largamente usato) per ottenere la tolleranza al guasto e' quello di eseguire computazioni multiple attraverso canali multipli. Quando si prevede la tolleranza ai guasti fisici, i canali possono essere identici, basandosi sull'ipotesi che i componenti hardware falliscono indipendentemente; un tale approccio non e' adatto per la tolleranza ai guasti di progetto dove i canali devono fornire *servizi identici* attraverso *progetti ed implementazioni distinti* [Elm 72, Ran 75, Avi 78], cioe' attraverso **diversita' di progetto (design diversity)**. [Avi 84].

Nel trattare sistemi tolleranti al guasto, si incontrano frequentemente situazioni che coinvolgono guasti e/o fallimenti multipli. Considerando le loro cause si possono distinguere:

- **guasti indipendenti (independent faults)**, che sono attribuiti a cause differenti,
- **guasti correlati (related faults)**, che sono attribuiti ad una causa comune.

Esempi di cause comuni sono alimentazioni (ove non replicate), orologi, specifiche, etc. Nella diversita' di progetto, i guasti correlati possono anche derivare dalle dipendenze fra progetti ed implementazioni distinti. I guasti correlati causano in genere **fallimenti a modo comune (common-mode failures)**.

Un'altra utile nozione quando si discute di sistemi tolleranti al guasto e' quella di **errori coincidenti (coincident errors)**, cioe' errori resi manifesti dallo stesso ingresso [Avi 86].

La relazione temporale fra fallimenti multipli porta a distinguere:

- **fallimenti simultanei (simultaneous failures)**, che si verificano entro una certa finestra temporale predefinita;
- **fallimenti sequenziali (sequential failures)**, che non si verificano entro la stessa finestra temporale predefinita.

Sebbene una definizione matematica del concetto di "simultaneo" porterebbe a zero la larghezza della finestra temporale, in pratica, tale finestra puo' essere piuttosto ampia, dipendendo dall'applicazione, e pertanto il termine di fallimenti "quasi coincidenti" viene talvolta usato [Tri 84]. Tipicamente, in un sistema tollerante al guasto che e' stato progettato per tollerare un guasto singolo alla volta, e' necessario recuperare dagli effetti di un guasto prima che il sistema possa tollerare il guasto successivo. In questo contesto, la finestra temporale che separa i guasti simultanei da quelli sequenziali e' l'intervallo di tempo necessario per il trattamento dell'errore e possibilmente il trattamento del guasto, durante il quale il sistema e' vulnerabile.

Un aspetto importante nella coordinazione delle attivita' di componenti multipli e' quella di evitare che la propagazione degli errori influenzi l'operazione di componenti non falliti. Questo aspetto diventa particolarmente importante quando un dato componente deve comunicare qualche informazione, che lui solo possiede, ad altri componenti. Esempi tipici di tale *informazione con singola sorgente* sono i sensori locali di dati, il valore di un orologio locale, la vista locale dello stato di altri componenti, etc. La conseguenza di questa necessita' di comunicazione di informazione con singola sorgente da un componente ad altri e' che i componenti non falliti devono raggiungere un *accordo* sul come usare l'informazione ottenuta in modo mutuamente consistente. Attenzione specifica a questo problema e' stata dedicata nel campo dei sistemi distribuiti (vedi ad esempio trasmissione per diffusione atomica [Cri 85a], la sincronizzazione degli orologi [Lam 85, Kop 87], o i protocolli di appartenenza [Cri 88]). E' importante capire comunque che la presenza inevitabile di ridondanza strutturale in un qualsiasi sistema tollerante al guasto implica distribuzione o ad un livello o ad un altro, e che il problema dell'accordo pertanto e' sempre presente. Sistemi tolleranti al guasto, geograficamente localizzati possono usare soluzioni al problema dell'accordo che sarebbero giudicate troppo costose in un sistema distribuito "classico" composto da componenti che comunicano tramite messaggi (ad esempio gli inter stadi [Lal 86], stadi multipli per la consistenza interattiva [Fri 82]).

La conoscenza di alcune proprieta' del sistema puo' limitare la quantita' di ridondanza necessaria, portando alla cosiddetta "tolleranza al guasto a basso costo". Esempi di queste proprieta' sono le regolarita' di natura strutturale: codici rilevatori e correttori di errore [Pet 72], strutture dati robuste [Tay 80], multiprocessori e reti di calcolatori [Pra 86, Ren 86], tolleranza al guasta basata su algoritmi [Hua 82]. I guasti che sono tollerati sono allora dipendenti dalle proprieta' che si tengono in conto, in quanto esse intervengono direttamente nelle ipotesi di guasto.

E' importante la segnalazione del fallimento di un componente ai suoi utenti. Di cio' si puo' tener conto nel contesto delle *eccezioni* [Mel 77, Cri 80, And 81]. La *gestione delle eccezioni (exception handling)* presente in alcuni linguaggi puo' costituire un modo conveniente per implementare il recupero dall'errore, in particolare il recupero in avanti[18].

(Anche) la tolleranza al guasto e' un concetto ricorsivo: e' essenziale che i meccanismi tendenti ad implementare la tolleranza al guasto siano protetti nei confronti dei guasti che li possono influenzare. Esempi di cio' sono la replicazione dei votatori, i controllori che si auto-controllano [Car 68], la memoria "stabile" per il recupero dei dati e dei programmi [Lam 81].

La tolleranza al guasto non e' limitata ai guasti accidentali. La protezione contro le intrusioni coinvolge tradizionalmente la crittografia [Den 82]. Alcuni meccanismi di rilevazione di errore sono orientati sia nei confronti di guasti intenzionali che accidentali (ad esempio le tecniche di protezione sugli accessi in memoria) e sono stati proposti schemi per la tolleranza sia alle intrusioni che ai guasti fisici [Fra 86, Rab 89], sia per la tolleranza alla logica maliziosa [Jos 88].

5.3 - Eliminazione del guasto

La eliminazione del guasto e' composta da tre fasi: verifica, diagnosi e correzione.

La **verifica (verification)** e' il processo di controllare se il sistema e' aderente alle proprieta', definite *condizioni di verifica* [Che 81]; se non e' aderente, si deve passare alle altre due fasi: diagnosticare i guasti che hanno impedito alle condizioni di verifica di essere rispettate, e poi eseguire le correzioni necessarie. Dopo la correzione, il processo deve riprendere per poter controllare che la eliminazione del guasto non abbia conseguenze

[18] L'uso del termine "eccezione", dovuto alla sua origine di trattare situazioni eccezionali - non solo errori - deve essere usato con attenzione nell'ambito della tolleranza al guasto: potrebbe apparire in contraddizione al considerare la tolleranza al guasto come attributo naturale dei sistemi di elaborazione, presa in considerazione dalle primissime fasi iniziali di progetto, e non un attributo "eccezionale".

indesiderate; la verifica eseguita a questo stadio e' chiamata di solito **verifica di** (non) **regressione ((non-) regression verification)**. Le condizioni di verifica possono assumere due forme:

- condizioni generali, che si applicano ad una data classe di sistemi, e sono pertanto - relativamente - indipendenti dalla specifica, ad esempio l'assenza di "abbraccio mortale" (deadlock), conformita' alle regole di progetto e di realizzazione;
- condizioni particolari per il sistema considerato, dedotte direttamente dalle sue specifiche.

Le tecniche di verifica possono essere classificate a seconda che esse comportino o no la messa in esercizio del sistema. Verificare un sistema senza effettiva esecuzione prende il nome di **verifica statica (static verification)**. La verifica puo' essere effettuata:

- sul sistema stesso, nella forma di a) *analisi statica* (ad esempio ispezioni o transitamenti (walk-through) [Mye 79], analisi del flusso dei dati [Ost 76], analisi di complessita' [McC 76], controlli a tempo di compilazione, etc.) o b) *prova di correttezza (proof-of-correctness)* (asserzioni induttive [Hoa 69, Cra 87]);
- su un modello di comportamento del sistema (ad esempio le reti di Petri, automi a stati finiti), portando ad una *analisi di comportamento* [Dia 82].

Verificare un sistema tramite la sua messa in esercizio costituisce la **verifica dinamica (dynamic verification)**; gli ingressi forniti al sistema possono essere sia simbolici nel caso della **esecuzione simbolica (symbolic execution)**, o con valori come nel caso del test di verifica, chiamato semplicemente **test (testing)**.

Il test esaustivo di un sistema rispetto a tutti i suoi ingressi possibili e' in genere non pratico. I metodi per la determinazione delle configurazioni di test possono essere classificati secondo due punti di vista: i criteri per la scelta degli ingressi di test, e la generazione degli ingressi di test.

I *criteri* per la scelta degli ingressi di test si possono considerare secondo tre punti di vista:

- lo scopo del test: il controllo se il sistema soddisfa le sue specifiche (funzionali) prende il nome di **test di conformita' (conformance testing)**, mentre il test teso alla rilevazione dei guasti si chiama **test di individuazione (fault-finding testing)**;
- il modello del sistema: a seconda se i criteri sono relativi alla funzione o alla struttura del sistema, essi portano rispettivamente al **test**

funzionale (functional testing) e al **test strutturale (structural testing)**;

- l'esistenza di un modello di guasto: se c'e' un tale modello di guasto, viene condotto il **test basato sul guasto (fault-based testing)** [Mor 90}, teso a rivelare particolari classi di guasti (ad esempio guasti a valore bloccato nella produzione dell'hardware [Rot 67], guasti fisici che influenzano il set di istruzioni di un microprocessore [Tha 78], guasti di progetto nel software [Goo 75, DeM 78, How 87]); se non esiste il modello di guasto, i criteri possono essere relativi ad esempio alla sensibilizzazione dei cammini [Rap 85, Nta 88], ai valori marginali di ingresso [Mye 79] nel software, etc.[19]

La *generazione* degli ingressi di test puo' essere deterministica o probabilistica:

- nel **test deterministico (deterministic testing)**, le configurazioni di test sono predeterminate da una scelta selettiva secondo i criteri adottati,
- nel **test casuale (random testing)**, o **statistico (statistical testing)**, le configurazioni di test sono scelte in accordo con una distribuzione di probabilita' definita rispetto al dominio degli ingressi; la distribuzione ed il numero dei dati di ingresso sono determinati in accordo ai criteri adottati [Dav 81, Dur 84].

La combinazione dei vari punti di vista porta ad un certo numero di approcci di test, alcuni dei quali vengono contrassegnati in forma sintetica, ad esempio:

- il test strutturale quando sia applicato all'hardware in genere significa test di individuazione, strutturale, basato sul guasto, mentre quando sia applicato al software significa test di individuazione, strutturale, non basato sul guasto;
- il test di *mutazione* del software [DeM 78] e' un test di individuazione, strutturale, deterministico, basato sul guasto.

Osservare le uscite del test e decidere se esse soddisfano o no le condizioni di verifica e' noto come problema dell'*oracolo* [Adr 82]. Le condizioni di verifica si possono applicare all'intero insieme delle uscite o ad una funzione compatta di queste ultime (ad esempio una segnatura di sistema quando si faccia il test per guasti fisici nell'hardware [Dav 86], o un "oracolo parziale"

---

19 La possibilita' di definire un modello di guasto e' strettamente correlata allo stadio nel processo di sviluppo che si considera: piu' avanzato e' lo stadio, piu' alta e' la possibilita' di definire un modello di guasto.

quando si faccia il test per guasti di progetto del software [Wey 82]). Quando si faccia il test per guasti fisici, i risultati - in forma compatta o no - anticipati dal sistema sotto test per una data sequenza di ingressi sono determinati per mezzo di simulazione [Lev 86] o per mezzo di un sistema di riferimento (golden unit). Per i guasti di progetto, il riferimento e' in generale la specifica; puo' anche essere un prototipo, o un'altra implementazione della stessa specifica nel caso di diversita' di progetto (test testa a testa (back-to-back testing), vedi ad esempio [Bis 88]).

Alcuni metodi di verifica possono essere usati insieme, ad esempio una esecuzione simbolica puo' essere usata a) per facilitare la determinazione delle configurazioni di test [Adr 82], o b) come metodo di prova di correttezza [Car 78].

Poiche' la verifica deve essere eseguita durante l'intero sviluppo del sistema, le tecniche precedenti si applicano naturalmente alle varie forme assunte dal sistema durante il suo sviluppo: prototipo, componente, etc. Verificare che il sistema non possa fare *di piu'* di cio' per cui e' specificato e' particolarmente importante rispetto ai guasti intenzionali [Gas 88].

Progettare un sistema in modo da facilitare la sua verifica e' cio' che prende il nome di **progetto per la verificabilita' (design for verifiability)**. Questo e' in particolare sviluppato per l'hardware rispetto ai guasti fisici, dove le tecniche corrispondenti prendono il nome di *progetto per la testabilita'* [Wil 83, McC 86].

La eliminazione del guasto durante la fase operativa della vita di un sistema e' la **manutenzione correttiva (corrective maintenance)**, tesa a preservare o migliorare la capacita' del sistema a fornire un servizio in accordo con la specifica[20]. La manutenzione correttiva puo' assumere due forme:

- manutenzione **curativa**, tesa a rimuovere guasti che abbiano prodotto uno o piu' errori e che siano stati registrati;
- manutenzione **preventiva**, tesa a rimuovere guasti prima che producano errori; questi guasti possono essere
  - guasti fisici che si siano verificati dalle ultime azioni di manutenzione preventiva,

---

20 Le altre forme di manutenzione usualmente distinte sono [Ram 84]:
- *manutenzione adattiva (adaptive maintenance)*, che adegua il sistema alle variazioni ambientali (ad esempio cambiamenti dei sistemi operativi o di basi di dati del sistema);
- *manutenzione perfettiva (perfective maintenance)*, che migliora la funzione del sistema in risposta a variazioni definite dal cliente - e dal progettista -, che possono coinvolgere la rimozione di guasti di specifica.

- guasti di progetto che abbiano portato ad errori in altri sistemi similari [Ada 84].

Queste definizioni[21] si applicano sia a sistemi non tolleranti il guasto che a sistemi tolleranti il guasto, che possono essere manutenibili in linea (senza interrompere la fornitura del servizio) o fuori linea. E' infine degno di nota che la frontiera fra manutenzione correttiva e trattamento del guasto e' relativamente arbitraria; in particolare, la manutenzione curativa puo' essere considerata come un mezzo - estremo - di ottenere tolleranza al guasto.

5.4 - Previsione del guasto

La previsione del guasto si effettua eseguendo una *valutazione* del comportamento del sistema rispetto all'occorrenza o alla attivazione del guasto. La valutazione ha due aspetti:

- *non probabilistico*, ad esempio determinare il cutset o il pathset minimo di un albero di guasto, conducendo una analisi dei modi di fallimento e dei loro effetti;
- *probabilistico*, teso a determinare la conformita' del sistema agli obiettivi di garanzia di funzionamento espressi in termini delle probabilita' associate ad alcuni degli attributi della garanzia di funzionamento, che possono allora essere definiti come *misure* della garanzia di funzionamento.

La vita di un sistema e' percepita dal suo utente come una alternanza fra due stati del servizio fornito rispetto alla specifica:

- **servizio corretto**, quando il servizio fornito *e' in accordo* con la specifica[22];
- **servizio non corretto**, quando il servizio fornito *non e' in accordo con* la specifica.

Un fallimento e' pertanto una transizione da un servizio corretto ad un servizio non corretto, e la transizione da un servizio non corretto ad uno

21 E' degno di nota che la discussione attuale circa l'irrilevanza dell'uso del termine "manutenzione" applicato al software dimentica semplicemente l'etimologia del termine: nel Medioevo, manutenzione designava le azioni eseguite per mantenere un esercito in condizioni di combattere, includendo percio' le forme correttiva, adattiva e perfettiva di manutenzione. L'associazione della manutenzione alla riparazione dell'hardware e' di fatto una (recente) deviazione; associare "manutenere" alla nozione di servizio permetterebbe di far rivivere questo significato etimologico, rimuovendo allo stesso tempo la causa di discussione.

22 Abbiamo deliberatamente ristretto l'uso del termine "corretto" al servizio fornito dal sistema, e non lo usiamo per il sistema stesso: secondo noi, difficilmente esistono sistemi non guasti, esistono soltanto sistemi che possono non avere ancora fallito.

corretto e' un **ripristino (restoration)**. Quantificare l'alternanza fra servizio fornito corretto e non corretto permette di definire l'affidabilita' e la disponibilita' come misure della garanzia di funzionamento :

- **affidabilita'**: una misura della fornitura *continua* di servizio corretto - o, analogamente, del *tempo al* fallimento;
- **disponibilita'**: una misura della fornitura di servizio corretto *rispetto all'alternanza* fra servizio corretto e non corretto.

Una terza misura, **manutenibilita'**, viene di solito considerata, ed puo' essere definita come una misura del tempo al ripristino dall'ultimo fallimento, o analogamente, della fornitura continua di servizio non corretto.

Come misura, la sicurezza di funzionamento puo' essere considerata come estensione della affidabilita'. Raggruppiamo lo stato del servizio corretto insieme con lo stato del servizio non corretto susseguente a fallimenti benigni in uno stato di funzionamento sicuro (nel senso di essere esente da danni catastrofici, non dal pericolo); la **sicurezza di funzionamento (safety)** e' allora una misura della sicurezza di funzionamento continua, o analogamente del tempo al fallimento catastrofico. La sicurezza di funzionamento puo' pertanto essere considerata come l'affidabilita' rispetto ai fallimenti catastrofici. Una estensione diretta della disponibilita', cioe' una misura della sicurezza di funzionamento rispetto all'alternanza fra stati di funzionamento sicuro e servizio non corretto dopo fallimenti catastrofici, non fornirebbe una misura significativa. Quando un fallimento catastrofico si e' verificato, le conseguenze sono di solito talmente importanti che il ripristino del servizio non e' di primaria importanza per almeno le due ragioni seguenti:

- e' secondario rispetto alla riparazione (in senso lato, includendo anche gli aspetti legali) delle conseguenze della catastrofe;
- il lungo periodo prima di permettere la ripresa delle operazioni (commissioni di inchiesta, etc.) porterebbe a valori numerici insignificanti.

Una misura "ibrida" del tipo affidabilita'-disponibilita' puo' comunque essere definita: una misura della fornitura del servizio corretto rispetto all'alternanza del servizio corretto e del servizio non corretto dopo fallimenti benigni. Questa misura e' di interesse in quanto fornisce una quantificazione della disponibilita' del sistema *prima* del verificarsi di un fallimento catastrofico, e come tale permette di quantificare il cosiddetto "compromesso affidabilita' (o disponibilita') e sicurezza di funzionamento".

Nel caso di sistemi ad operazioni multiple, possono essere distinti diversi servizi, cosi' come diversi modi di fornitura del servizio, con una gamma che va dalla piena operativita' alla completa inoperativita', che puo' essere vista come caratterizzante forniture di servizi sempre meno corrette. Misure di garanzia di funzionamento relative all'operativita' per tali sistemi sono di solito definite di **prestazione-affidabilita' congiunte (performability)** [Mey 78, Smi 88].

I due principali approcci alla previsione del guasto probabilistica, tesi a derivare stime quantitative delle misure della garanzia di funzionamento, sono il modellamento ed il test (di valutazione). Questi approcci sono direttamente complementari, poiche' il modellamento necessita di dati relativi ai processi base modellati (processo di fallimento, processo di manutenzione, processo di attivazione del sistema, etc.), che possono essere ottenuti per mezzo del test.

Quando si esegua una valutazione attraverso *modellamento (modeling)*, gli approcci differiscono significativamente a seconda se il sistema e' considerato essere in affidabilita' stabile o in crescita di affidabilita', che possono essere definite come segue [Lap 90b]:

- **affidabilita' stabile (stable reliability)**: la capacita' del sistema di fornire servizio corretto e' *preservata* (identita' stocastica dei tempi al fallimento successivi);
- **crescita di affidabilita' (reliability growth)**: la capacita' del sistema di fornire servizio corretto e' *aumentata* (incremento stocastico dei tempi al fallimento successivi)[23].

Interpretazioni pratiche della affidabilita' stabile e della crescita di affidabilita' sono le seguenti:

- affidabilita' stabile: ad un dato ripristino, il sistema e' identico a quello precedente il ripristino; questo corrisponde alle seguenti situazioni:
  - nel caso di un fallimento hardware, la parte fallita e' sostituita da un'altra, identica e non fallita,
  - nel caso di fallimento software, il sistema e' fatto ripartire con una configurazione di ingresso differente da quella che ha portato al fallimento;
- crescita di affidabilita': il guasto la cui attivazione ha portato al fallimento e' diagnosticato come guasto di progetto (dell'hardware o del sofware) ed e' rimosso.

---

23 *Diminuzione di affidabilita' (reliability decrease)* (la capacita' del sistema di fornire servizio corretto e' diminuita, e c'e' pertanto un decremento stocastico dei tempi al fallimento successivi) e' teoricamente, e praticamente, possibile, ad esempio con l'introduzione di nuovi guasti durante una azione correttiva, la cui probabilita' di attivazione e' piu' grande della probabilita' di attivazione dei guasti rimossi. In tal caso, si deve sperare che la diminuzione sia limitata nel tempo, e che l'affidabilita' cresca globalmente su un periodo di tempo di osservazione prolungato.

La valutazione della garanzia di funzionamento dei sistemi in affidabilita' stabile e' di solito composta di due fasi principali:

- *costruzione* del modello del sistema a partire dai processi stocastici elementari che modellano il comportamento dei componenti del sistema e le loro interazioni;
- *elaborazione* del modello per poter ottenere le espressioni ed i valori delle misure di garanzia di funzionamento del sistema.

La valutazione puo' essere fatta rispetto a a) guasti fisici [Tri 84], b) guasti di progetto [Lit 79, Arl 88], o c) una combinazione di essi [Lap 84, Pig 88]. La garanzia di funzionamento di un sistema e' fortemente dipendente dal suo ambiente, nel senso piu' ampio del termine [Hec 87], o piu' specificamente dal suo carico [Cas 81, Iye 82].

Molti modelli di crescita di affidabilita' sono stati proposti, per l'hardware [Dua 64], per il software, o entrambi [Lap 90]. La maggior parte di essi sono per il software, e sono orientati a valutare o l'affidabilita' [Yam 85, Mil 86], o il numero dei guasti (rimanenti) [Goe 79, Toh 89]; siccome questi modelli sono orientati a predire l'affidabilita' futura a partire dai dati di fallimento accumulati nel passato, particolare attenzione e' stata dedicata al problema della predizione [Lit 88].

Sebbene non sia principalmente teso a verificare un sistema, il *test di valutazione (evaluation testing)* [Mil 87, Cho 87] puo' essere caratterizzato usando i punti di vista definiti nel capitolo 5.3: conformita', funzionale, non basato sul guasto, statistico. La preoccupazione principale in questo campo e' che il profilo di ingresso sia rappresentativo del profilo operativo, da cui il nome: **test operazionale (operational testing)**.

Quando si valutano sistemi a funzionamento garantito, la copertura dei meccanismi di trattamento dell'errore e del trattamento del guasto ha una influenza molto significativa [Bou 69, Arn 73]; la sua valutazione puo' essere effettuata o tramite modellamento [Dug 89] o tramite test, chiamato allora *iniezione del guasto (fault-injection)* [Arl 89, Gun 89].

6 - GLI ATTRIBUTI DELLA GARANZIA DI FUNZIONAMENTO

Gli attributi sono stati definiti nel capitolo 1 in accordo a proprieta' differenti, che possono essere piu' o meno messe in evidenza a seconda dell'applicazione del sistema di elaborazione che si sta considerando:

- la disponibilita' e' sempre richiesta, sebbene ad un livello che dipende dall'applicazione;
- l'affidabilita', la sicurezza di funzionamento, la sicurezza-confidenzialita' possono o no essere richieste a seconda dell'applicazione.

Una proprieta' addizionale, che puo' essere vista come prerequisito per l'ottenimento delle altre proprieta', e' l'**integrita' (integrity)**, cioe' la condizione di non essere deteriorato, nel senso piu' ampio del termine: a) sia per i dati che per il programma, e b) rispetto a guasti accidentali o intenzionali.

Le variazioni sull'enfasi da porre sugli attributi della garanzia di funzionamento hanno una influenza diretta sul giusto bilanciamento delle tecniche descritte nel capitolo precedente da usare perche' il sistema risultante sia a funzionamento garantito. Questo e' fra tutti il problema piu' difficile poiche' alcuni degli attributi sono antagonisti (ad esempio disponibilita' e sicurezza di funzionamento, disponibilita' e sicurezza-confidenzialita'), avendo bisogno di compromessi per essere attuati. Considerando le tre dimensioni principali di progetto di un sistema di elaborazione, cioe' il costo, la prestazione e la garanzia di funzionamento, il problema e' reso ancora piu' acuto dal fatto che la dimensione della garanzia di funzionamento e' meno dominata dello spazio di progetto costo-prestazione [Sie 82].

La manutenibilita' e' stata definita nel paragrafo 5.4 come una misura della garanzia di funzionamento. La manutenibilita' puo' essere anche considerata come un attributo di garanzia di funzionamento, relativo alla facilita' con cui le azioni di manutenzione possono essere eseguite.

La definizione di sicurezza-confidenzialita' data nel capitolo 1, prevenzione degli accessi non autorizzati e/o gestione non autorizzata dell'informazione, origina da [ITS 90], che definisce la sicurezza-confidenzialita' come la combinazione della confidenzialita', prevenzione della divulgazione non autorizzata di informazione, della integrita', prevenzione della cancellazione o variazione non autorizzate di informazione, e della disponibilita', prevenzione del trattenimento non autorizzato di informazione. Si noti che:

a) un accesso o gestione di informazione non autorizzati puo' derivare sia da guasto accidentale che intenzionale, e come osservato nel paragrafo 5.2, alcuni meccanismi per proteggere contro gli accessi non autorizzati sono comuni ad entrambi i tipi di guasto;
b) rispetto ai guasti intenzionali, la nozione di autorizzazione deve essere intesa in senso lato: una persona autorizzata che abusa della sua

autorita' di fatto viola, eseguendo azioni illegittime, l'autorizzazione che le era stata concessa.

Una caratteristica importante dei sistemi tolleranti il guasto e' la loro capacita', a causa della presenza delle procedure di rilevazione dell'errore, di fornire agli utenti l'informazione se il servizio (funzionale) che viene fornito e' corretto o no. Tale proprieta' e' definita **attendibilita' (trustability)** in [Fur 90].

La *valutazione* se un sistema e' veramente a funzionamento garantito - fiducia giustificata sul servizio fornito - o no va pertanto oltre le tecniche di validazione cosi' come sono state descritte nel capitolo precedente per almeno le tre seguenti ragioni e limitazioni:

- controllare con certezza la copertura del progetto o le ipotesi di validazione rispetto alla realta' (ad esempio la corrispondenza ai guasti effettivi dei criteri usati per determinare gli ingressi di test, le ipotesi di guasto nel progetto dei meccanismi di tolleranza al guasto) implicherebbe una conoscenza ed un dominio della tecnologia usata, dell'uso inteso del sistema, etc. che sono ben oltre a cio' che e' ottenibile in generale;
- eseguire una valutazione di un sistema secondo alcuni attributi di garanzia di funzionamento rispetto a qualche classe di guasto e' attualmente considerato non possibile o non significativo: le basi teoriche probabilistiche non esistono o non sono - ancora - largamente accettate; esempi sono la sicurezza di funzionamento rispetto a guasti accidentali di progetto, la sicurezza-confidenzialita' rispetto a guasti intenzionali;
- le specifiche "contro" le quali viene eseguita la validazione non sono in generale esenti da guasti - come qualsiasi sistema.

Fra le numerose conseguenze di questo stato di cose, citiamo:

- l'enfasi posta sul processo di sviluppo e di produzione quando si valuti un sistema: metodi e tecniche utilizzate e come sono usate; in alcuni casi, un *voto* e' assegnato e fornito al sistema secondo a) la natura dei metodi e delle tecniche usati, e b) una valutazione della loro utilizzazione[24];

---

[24] Ad esempio:
- i sistemi sono classificati da un punto di vista di sicurezza [DoD 85] da A1 ("progetto verificato") a D ("protezione minima");
- il software per aeroplani per trasporto covile sono classificati [RTC 85] come Livello 1, 2 o 3 a seconda della criticita' delle funzioni che devono essere espletate dal software: critiche, essenziali, non essenziali.

- la presenza nella specifica di qualche sistema tollerante il guasto, in aggiunta ai requisiti probabilistici in termini di misure di garanzia di funzionamento, del numero dei guasti che devono essere tollerati[25]; tale specifica non sarebbe necessaria se le limitazioni menzionate potessero essere superate.

CONCLUSIONI

Sempre di piu', individui ed organizzazioni stanno sviluppando o procurandosi sistemi di elaborazione sofisticati sui cui servizi devono riporre grande fiducia - sia per servire distributori automatici di denaro, per calcolare orbite di satelliti, per controllare un aeroplano o un impianto nucleare, e per mantenere la confidenzialita' di una base di dati riservata. In circostanze differenti, l'attenzione sarebbe posta su differenti proprieta' di tali servizi - ad esempio il tempo medio di risposta ottenuto, la probabilita' di produrre i risultati richiesti, la capacita' di evitare fallimenti catastrofici per l'ambiente del sistema, o il grado entro cui prevenire intrusioni intenzionali. La nozione di garanzia di funzionamento fornisce un mezzo molto conveniente per includere questi vari interessi entro un quadro concettuale unico. La garanzia di funzionamento pertanto include come casi particolari proprieta' come l'affidabilita', la disponibilita', la sicurezza di funzionamento, la sicurezza-confidenzialita'. Essa fornisce anche un mezzo per affrontare il problema di cio' che un utente in genere ha bisogno da un sistema, e cioe' di un *bilanciamento appropriato* di tali proprieta'

---

[25] Tali specifiche sono classiche nelle applicazioni aerospaziali, sotto la forma di una concatenazione di requisiti di "fallimento operativo" o "fallimento non pericoloso".

GLOSSARIO

Attenzione: questo glossario e' fornito come aiuto per la lettura del documento. Non va considerato indipendentemente dal documento stesso.

Affidabilita'................................Garanzia di funzionamento rispetto alla continuita' del servizio.
Misura della fornitura di servizio continuo corretto.
Misura del tempo al fallimento.

Affidabilita' stabile....................La capacita' del sistema di fornire servizio corretto e' conservata (identita' stocastica dei tempi al fallimento successivi).

Ambiente del sistema.................Gli altri sistemi che interagiscono o interferiscono con il sistema dato.

Attendibilita'.............................La capacita' del sistema di fornire agli utenti informazioni circa la correttezza del servizio.

Attributi della garanzia di funzionamento........Attributi che permettono la valutazione della qualita' del sistema risultante dagli impedimenti e dai mezzi che li si oppongono. Affidabilita', disponibilita', manutenibilita', sicurezza di funzionamento, sicurezza-confidenzialita', attendibilita'.

Compensazione dell'errore........Forma di trattamento di errore quando lo stato erroneo contiene sufficiente informazione (ridondanza) per permettere la fornitura di servizio corretto.

Componente che si autocontrolla.............................Componente che include meccanismi di rilevazione di errore associati alle sue parti funzionali.

Componente del sistema ....Un altro sistema.

Comportamento del sistema.......Cio' che un sistema fa.

Conseguimento della garanzia di funzionamento .... Metodi e tecniche intese a dotare un sistema della capacita' di fornire un servizio in accordo con la specifica.

Prevenzione del guasto e tolleranza al guasto.

Copertura .... Misura della rappresentativita' delle situazioni a cui un sistema e' sottoposto durante la sua validazione in confronto a quelle effettive durante la sua vita operativa.

Crescita di affidabilita' .... La capacita' del sistema di fornire servizio corretto e' aumentata (incremento stocastico dei tempi al fallimento successivi.

Criticita' del sistema .... La piu' alta gravita' dei modi di fallimento.

Diagnosi di guasto .... L'azione di determinare la causa di un errore sia come posizione che come natura.

Disattivazione del guasto .... Le azioni intraprese perche' un guasto non possa essere attivato.

Disponibilita' .... Garanzia di funzionamento rispetto alla prontezza all'uso.

Misura della fornitura di servizio corretto rispetto all'alternanza di servizio corretto e non corretto.

Diversita' di progetto .... Un approccio alla produzione di sistemi, che comporta la fornitura di servizi identici da progetti ed implementazioni differenti.

Eliminazione del guasto .... Metodi e tecniche tese a ridurre la presenza (numero, gravita') dei guasti.

Errore .... Parte dello stato del sistema che e' responsabile di condurre ad un fallimento.

Errore latente .... Errore non riconosciuto come tale.

Errore rilevato .... Errore riconosciuto come tale da un algoritmo o meccanismo di rilevazione.

Errori coincidenti .... Errori prodotti dallo stesso ingresso.

Esecuzione simbolica .... Verifica dinamica eseguita con ingressi simbolici.

Fallimenti a modo comune........Fallimenti che derivano da guasti correlati.

Fallimenti sequenziali................Fallimenti che non si verificano entro la stessa finestra temporale predefinita.

Fallimenti simultanei................Fallimenti che si verificano entro qualche finestra temporale predefinita.

Fallimento o

Malfunzionamento......................Deviazione del servizio fornito rispetto alla specifica.

Transizione da fornitura di servizio corretto a fornitura di servizio non corretto.

Fallimento arbitrario.................vedi Fallimento.

Fallimento benigno....................Fallimento le cui conseguenze sono dello stesso ordine di grandezza del beneficio prodotto dalla fornitura di servizio corretto.

Fallimento catastrofico..............Fallimento le cui conseguenze sono incommensurabilmente piu' grandi del beneficio prodotto dalla fornitura di servizio corretto.

Fallimento con blocco................L'attivita' del sistema, se c'e', non e' piu' percepibile all'utente, ed un servizio a valore costante e' fornito.

Fallimento consistente...............Fallimento percepito allo stesso modo da tutti gli utenti del sistema.

Fallimento inconsistente............Fallimento tale che gli utenti del sistema possono avere percezioni differenti di esso.

Fallimento nel tempo.................Fallimento tale che la temporizzazione della fornitura di servizio non e' in accordo alla specifica.

Fallimento nel valore.................Fallimento tale che il valore del servizio fornito non e' in accordo con la specifica.

Fallimento per omissione

persistente...................................Fallimento per omissione persistente.

Fallimento per omissione..........Fallimento tale che nessun servizio e' fornito.

Funzione del sistema.................Cio' che un sistema e' inteso fare.

Funzione in tempo reale............Funzione che deve essere eseguita entro intervalli di tempo finiti dettati dall'ambiente.

Garanzia di funzionamento........Attendibilita' di un sistema di elaborazione tale che affidamento possa essere giustificatamente riposto sul servizio che esso fornisce.

Gravita' del fallimento..............Livello delle conseguenze del fallimento sull'ambiente del sistema.

Guasti correlati..........................Guasti attribuiti ad una causa comune.

Guasti indipendenti....................Guasti attribuiti a cause differenti.

Guasto..Causa aggiudicata o ipotizzata di un errore.

Causa di errore che si intende evitare o tollerare.

Conseguenza per un sistema del fallimento di un altro sistema che ha interagito o sta interagendo con il sistema che si considera.

Guasto accidentale......................Guasto che si verifica o e' creato fortuitamente.

Guasto attivo..............................Guasto che produce un errore.

Guasto causato dall'uomo..........Guasto che deriva da imperfezione umana.

Guasto di progetto......................Guasto interno ad un sistema causato dall'uomo.

Guasto esterno............................Guasto che deriva da interferenza o interazione ambientale.

Guasto fisico...............................Guasto che deriva da fenomeni fisici avversi.

Guasto forte................................Guasto che necessita di disattivazione.

Guasto inattivo............................Guasto interno non attivato dal processo di elaborazione.

Guasto intenzionale.....................Guasto creato deliberatamente.

Guasto intermittente...................Guasto interno temporaneo.

Guasti le cui condizioni di attivazione non possono essere riprodotte o che si verificano abbastanza raramente.

Guasto interno............................Guasto entro un sistema.

Guasto debole.............................Guasto che non necessita di disattivazione.

Guasto non ricorrente.................vedi guasto intermittente.

Guasto operativo........................Guasto che si verifica durante l'uso del sistema.

Guasto permanente......................Guasto la cui presenza non e' in relazione a condizioni puntuali temporalmente del sistema, sia interne che esterne.

Guasto ricorrente........................Guasto permanente le cui condizioni di attivazione possono essere riprodotte.

Guasto solido...............................vedi guasto forte.

Guasto temporaneo.....................Guasto che e' presente per un periodo limitato di tempo.

Guasto transitorio.......................Guasto esterno fisico temporaneo.

Impedimenti alla garanzia di funzionamento........Circostanze indesiderate, ma non inattese, che causano o che risultano da non garanzia di funzionamento.

Guasti, errori e fallimenti.

Integrita'.......................................Condizione di non essere impedito.

Intrusione.....................................Guasto operativo esterno intenzionale.

Logica maliziosa.........................Guasto di progetto intenzionale.

Manutenibilita'...........................Misura della fornitura continua di servizio non corretto.

Misura del tempo necessario per il ripristino a partire dall'ultimo fallimento verificato.

Manutenzione correttiva............Preservazione o miglioramento durante la sua vita operativa della capacita' del sistema di fornire un servizio in accordo con la specifica.

Eliminazione dei guasti durante la vita operativa del sistema.

Manutenzione curativa...............Manutenzione correttiva tesa a rimuovere i guasti che hanno prodotto errori e che sono stati evidenziati.

Manutenzione preventiva...........Manutenzione correttiva tesa a rimuovere i guasti prima che siano attivati.

Mascheramento del guasto.........Il risultato dell'applicazione sistematica della compensazione dell'errore, anche in assenza di errori.

Mezzi per ottenere la garanzia di funzionamento........Metodi e tecniche che permettono di a) dotare un sistema della capacita' di fornire un servizio su cui si possa porre fiducia, e b) raggiungere confidenza in questa capacita'.

Prevenzione del guasto, tolleranza al guasto, eliminazione del guasto, previsione del guasto.

Misure della garanzia di funzionamento........Attributi che permettono di stimare la qualita' del servizio che risulta dagli impedimenti e dai mezzi che li si oppongono. Affidabilita', disponibilita', manutenibilita', sicurezza di funzionamento.

Modo di evitare il guasto..........Metodi e tecniche tese a produrre un sistema esente da guasto.

Prevenzione del guasto e eliminazione del guasto.

Non garanzia di funzionamento............................Proprieta' di un sistema di elaborazione tale che fiducia non possa piu' essere, o non verra' piu', giustificatamente riposta sul servizio che esso fornisce.

Prestazione-affidabilita' congiunte....................................Misure di garanzia di funzionamento in relazione alla prestazione.

Prevenzione del guasto..............Metodi e tecniche tese a prevenire il verificarsi o l'introduzione del guasto.

Previsione del guasto.................Metodi e tecniche tese a valutare il numero attuale, l'incidenza futura, e le conseguenze dei guasti.

Progetto per verificabilita'.......Metodi e tecniche adottate nel progetto di un sistema che facilitino la sua verifica.

Punto di recupero......................Punto temporale durante l'esecuzione di un processo tale che lo stato corrente ad esso corrispondente possa dover essere successivamente ripristinato.

Recupero dall'errore.................Forma di trattamento dell'errore dove uno stato esente da errore e' sostituito al posto di uno stato erroneo.

Recupero in avanti........................Forma di recupero dall'errore dove la trasformazione dello stato erroneo consiste nel trovare un nuovo stato.

Recupero indietro........................Forma di recupero dall'errore dove la trasformazione dello stato erroneo consiste nel riportare indietro il sistema ad uno stato precedentemente occupato.

Rilevazione dell'errore..............L'azione di identificare che uno stato del sistema e' erroneo.

Ripristino del servizio...............Transizione da fornitura di servizio non corretto a fornitura di servizio corretto.

Servizio..Comportamento del sistema cosi' come percepito dall'utente.

Servizio corretto........................Servizio fornito in accordo con la specifica del sistema.

Servizio in tempo reale.............Servizio che deve essere fornito entro intervalli di tempo finiti dettati dall'ambiente.

Servizio non corretto.................Servizio fornito non in accordo con la specifica del sistema.

Sicurezza-confidenzialita'.........Garanzia di funzionamento rispetto alla prevenzione di accessi non autorizzati e/o gestione non autorizzata di informazione.

Sicurezza di funzionamento.......Garanzia di funzionamento rispetto al non verificarsi di guasti catastrofici.

Misura della fornitura continua sia di servizio corretto che di servizio non corretto dopo fallimenti benigni.

Misura probabilistica del tempo al fallimento catastrofico.

Sistema...Entita' che ha interagito, interagisce, o capace di interagire con altre entita'.

Insieme di componenti collegati insieme per interagire.

Sistema a fallimento con blocco..................................Sistema i cui fallimenti possono essere soltanto, o entro limiti accettabili, fallimenti con blocco.

Sistema a fallimento non pericoloso............................Sistema i cui fallimenti possono essere soltanto, o entro limiti accettabili, fallimenti benigni.

Sistema a fallimento tacito........ Sistema i cui fallimenti possono essere soltanto, o entro limiti accettabili, fallimenti per omissione persistente.

Sistema atomico........................ Sistema la cui struttura interna non puo' essere distinta, o non e' di interesse e puo' essere ignorata.

Sistema in tempo reale............... Sistema che esegue almeno una funzione in tempo reale o che fornisce almeno un servizio in tempo reale.

Specifica del sistema................. Descrizione concordata dei requisiti del sistema.

Stato del sistema........................ Una condizione di essere del sistema rispetto ad un insieme di circostanze.

Struttura del sistema.................. Cio' che fa si' che un sistema faccia cio' che fa.

Test.. Verifica dinamica eseguita con ingressi aventi valore appositamente predeterminato e/o scelto.

Test basato sul guasto................ Test teso a rivelare specifiche classi di guasto.

Test casuale................................ vedi test statistico.

Test deterministico.................... Forma di test in cui le configurazioni di test sono predeterminate da una scelta selettiva.

Test di conformita'.................... Test il cui scopo e' di controllare se il sistema soddisfa la sua specifica.

Test di identificazione del guasto.................................. Test il cui scopo e' rivelare guasti.

Test funzionale.......................... Forma di test dove gli ingressi di test sono scelti secondo criteri relativi alla funzione del sistema.

Test operazionale....................... Test eseguito per valutare la garanzia di funzionamento di un sistema, con un profilo di ingresso rappresentativo delle sue condizioni operative.

Test statistico............................ Forma di test dove le configurazioni di test sono scelte in base ad una distribuzione di probabilita' definita sul dominio degli ingressi.

Test strutturale.......................... Forma di test dove gli ingressi di test sono scelti in accordo a criteri relativi alla struttura del sistema.

Tolleranza al guasto..................Metodi e tecniche tese a fornire un servizio in accordo con la specifica nonostante i guasti.

Trattamento del guasto..............Le azioni intraprese per prevenire la riattivazione di un guasto.

Trattamento dell'errore.............Le azioni intraprese per eliminare gli errori da un sistema.

Utente del sistema......................Un altro sistema (fisico, umano) che interagisce con il sistema che si considera.

Validazione..................................Metodi e tecniche per raggiungere la fiducia nella capacita' di un sistema di fornire un servizio in accordo con la specifica.

Eliminazione e previsione del guasto.

Verifica......................................Il processo di determinare se un sistema e' congruente alle caratteristiche (le condizioni di verifica) che possono essere a) generali, indipendenti dalla specifica, o b) particolari, dedotte dalla specifica.

Verifica dinamica.......................Verifica che comporta la messa in esercizio del sistema.

Verifica di regressione..............Verifica eseguita dopo una correzione, per controllare che la correzione non ha conseguenze indesiderate.

Verifica statica...........................Verifica condotta senza la messa in esercizio del sistema.

INDICE ITALIANO - INGLESE

Affidabilita'..Reliability
Affidabilita' stabile....................................Stable reliability
Ambiente del sistema.................................System environment
Attendibilita'...Trustability
Attributi della garanzia di funzionamento ....Dependability attributes
Compensazione dell'errore..........................Error compensation
Componente che si autocontrolla.................Self-checking component
Componente del sistema.............................System component
Comportamento del sistema.........................System behavior
Conseguimento della garanzia
di funzionamento.......................................Dependability procurement

Copertura .... Coverage
Crescita di affidabilita' .... Reliability growth
Criticita' del sistema .... System criticality
Diagnosi di guasto .... Fault diagnosis
Disattivazione del guasto .... Fault passivation
Disponibilita' .... Availability
Diversita' del progetto .... Design diversity
Eliminazione del guasto .... Fault removal
Errore .... Error
Errori coincidenti .... Coincident errors
Errore latente .... Latent error
Errore rilevato .... Detected error
Esecuzione simbolica .... Symbolic execution
Fallimento .... Failure
Fallimenti a modo comune .... Common-mode failures
Fallimento arbitrario .... Arbitrary failure
Fallimento benigno .... Benign failure
Fallimento catastrofico .... Catastophic failure
Fallimento con blocco .... Stopping failure
Fallimento consistente .... Consistent failure
Fallimento inconsistente .... Inconsistent failure
Fallimento nel tempo .... Timing failure
Fallimento nel valore .... Value failure
Fallimento per omissione .... Omission failure
Fallimento per omissione persistente .... Crash failure
Fallimenti sequenziali .... Sequential failures
Fallimenti simultanei .... Simultaneous failures
Funzione del sistema .... System function
Funzione in tempo reale .... Real-time function
Garanzia di funzionamento .... Dependability
Gravita' del fallimento .... Failure severity
Guasto .... Fault
Guasto accidentale .... Accidental fault
Guasto attivo .... Active fault
Guasto causato dall'uomo .... Human-made fault
Guasti correlati .... Related faults
Guasto debole .... Soft fault
Guasto di progetto .... Design fault
Guasto esterno .... External fault

Guasto fisico..Physical fault
Guasto forte..Hard fault
Guasto inattivo..Dormant fault
Guasti indipendenti...................................Independent faults
Guasto intenzionale..................................Intentional fault
Guasto intermittente.................................Intermitent fault
Guasto interno..Internal fault
Guasto operativo.....................................Operational fault
Guasto permanente..................................Permanent fault
Guasto solido...Solid fault
Guasto temporaneo..................................Temporary fault
Guasto transitorio....................................Temporary fault
Impedimenti alla garanzia
di funzionamento....................................Impairments to dependability
Integrita'...Integrity
Intrusione..Intrusion
Logica maliziosa.....................................Malicious logic
Manutenzionabilita'.................................Maintenability
Manutenzione correttiva...........................Corrective maintenance
Manutenzione curativa.............................Curative maintenance
Manutenzione preventiva..........................Preventive fault
Mascheramento del guasto........................Fault masking
Mezzi per la garanzia di funzionamento.......Means for dependability
Modo di evitare il guasto..........................Fault avoidance
Prestazione-affidabilita' congiunte..............Performability
Prevenzione del guasto.............................Fault prevention
Previsione del guasto...............................Fault forecasting
Progetto per verificabilita'........................Design for verifiability
Punto di recupero....................................Recovery point
Recupero dall'errore.................................Error recovery
Recupero in avanti...................................Forward recovery
Recupero indietro....................................Backward recovery
Rilevazione dell'errore..............................Error detection
Ripristino del servizio..............................Service restoration
Servizio..Service
Servizio corretto......................................Correct service
Servizio in tempo reale.............................Real-time service
Servizio non corretto................................Incorrect service
Sicurezza di funzionamento.......................Safety

Sicurezza-confidenzialita'..........................Security
Sistema..System
Sistema a fallimento con blocco ....Fail-stop system
Sistema a fallimento non pericoloso.............Fail-safe system
Sistema a fallimento tacito..........................Fail-silent system
Sistema atomico ....Atomic system
Sistema in tempo reale...............................Real-time system
Specifica del sistema..................................System specification
Stato del sistema...System state
Struttura del sistema..................................System structure
Test ....Testing
Test basato sul guasto................................Fault-based testing
Test casuale ....Random testing
Test deterministico ....Deterministic testing
Test di conformita'....................................Conformance testing
Test di identificazione del guasto................Fault-finding testing
Test funzionale..Functional testing
Test operazionale ....Operational testing
Test statistico ....Statistic testing
Test strutturale..Structural testing
Tolleranza al guasto...................................Fault tolerance
Trattamento del guasto...............................Fault treatment
Trattamento dell'errore ....Error processing
Utente ....User
Validazione...Validation
Verifica..Verification
Verifica dinamica.......................................Dynamic verification
Verifica di regressione...............................Regression verification
Verifica statica..Static verification

デペンダビリティ：

概念と関連用語

はじめに

この記録文書は、コンピュータシステムのデペンダビリティに関する様々な属性を明確にするため、非公式ではあるが厳密に定義を行うことを目的としたものである。ここで示す内容は高信頼及びフォールトトレラントコンピューティングに関する科学及び技術研究部会(Avi 67, Jes 77, Mel 77, Avi 78, Ran 78, Car 79, And 81, FTC 82, Sie 82, Cri 85a, Lap 85, Avi 86, Lap 89)において行われた研究をもとに、できる限り明白にしかも広範囲にわたり受け入れられる基本的な定義を提案するものである。

まず最初にデペンダビリティの一般概念を明らかにする。これは、信頼性・アベイラビリティ・安全性・セキュリティ等システム利用者側から見た普遍的な属性と見なすことができる。第1章では、基本的な定義をその概要と共に与え、引き続く章ではその解説、追加すべき定義について述べる。本文書中に定義として与えた用語とその概略の意味を付録に示してある。記述にあたっては前に戻って参照することのないような構成を行っている。定義を行う用語については**強調した文字**で表し、特に注目してほしい用語は網目かけ文字で示してある。

ここで記述を行う方針について、次にその概要を示す。

- デペンダビリティの属性を表現する事ができる最も基本的な概念を明らかにする。
- 用語については可能ならば一般的に用いられている用語と同じに、あるいはできる限り同意に用いる。定義されていない用語は、普通に用いられている意味（辞典に示されている意味）にとどめるようにする。
- フォールトの分類については独立した定義を与えることにより、統合化に重点を置く［Gol 82, Ran 86 ］ことにする。

この記録文書は研究集会を通じて互いに意見の交換を行いながら得られた結実として、最小限得られたコンセンサスを基にしている。このことからここに示す記録文書は、ａ）他の組織体（規格化を行う機関を含む），ｂ）教育目的などにおける利用に適しているものと思われる。この様な観点から関連用語に関する検討はこれで終了したわけではなく、考え方を他の人々に伝えるための言葉であり、多くの批判を受けながら、多方面から寄せられる見解を基にして、より洗練されたものとしたい。この記録文書の目的は、現状技術を表す道標である。提示する概念は、その関連する技術および仕様内容に関する理解能力の涵養と共に、信頼しうるにたるコンピュータ・システムの設計とそのアセスメントの発展に寄与しうるものと確信する。

この記録文書が出版できる背景として、多くの方々から寄せられた多大な議論、特にIFIP WG 10.4の各位による貢献がなければ実現しなかったことを記しておきたい。

1. 基本定義

デペンダビリティ（dependability ）とはコンピュータシステムが与えられた仕事（service ）を実行するにあたり正確な信頼のおけるその資質と定義する［Car 82］。システムにより実行される**仕事**とは利用者によって認識されるシステムの動作であり、**利用者**はシステムと相互に作用する別のシステム（人及び物）とみなすことができる。

コンピュータシステムの応用目的によって、デペンダビリティに対する重点指向はそれぞれ異なった側面があろう。すなわちデペンダビリティは見方によって異なった様相を持ち、しかも相い補う**性質**（properties）であって、デペンダビリティの**属性**（attributes）を定義することができる。

- 即利用可能性（readiness for usage ）を必要とする場合には、その意味は**アベイラビリティ**（availability）である。
- 仕事の継続性（continuity of service ）を必要とする場合には、その意味は**信頼度**（reliability ）である。
- 周囲の環境に致命的な結果をもたらすことのないようこれを回避（avoidance of catastrophic consequences on the environment）する必要がある場合は、その意味は**安全度**である。
- 無許可で情報をアクセスしたり取り扱うことを防止（prevention of unauthorized access and/or handling information ）する場合は、**セキュリティ**（security）である。

システムの障害（failure ）は、実行される仕事が実行すべき仕事の仕様（specification）から逸脱しているとき生ずる。仕様はシステムの期待すべき機能及び仕事についての合意文書である。誤り（error)は障害をもたらすシステムの状態の一部であって、仕事に影響を与える誤りがあれば、それにより障害が生じる、あるいは障害が発生したという。誤りのはっきりした、あるいはその様に想定される原因がフォールト（fault ）である。

コンピュータシステムのデペンダビリティは、次にあげる方法を組み合せることよって向上をはかることができる。

- **フォールト予防**（fault prevention）：フォールトの発生（occurrence）及び生起（introduction）を予防する方法.
- **フォールトトレランス**（fault tolerance ）：フォールトが生じても仕様に示された仕事の遂行を可能とする方法.
- **フォールト除去**（fault removal ）：フォールトの存在（数，程度）を減少させる方法.
- **フォールト予測**（fault forecasting ）：フォールトの現存数・将来の影響・フォールト誤りによる結果を予測する方法.

フォールト予防とフォールトトレランスは、仕様に定められた仕事を実行する能力（ability ）をシステムに与える方法であり、デペンダビリティの**達成**（dependability procurement）とみなされる。

さらにフォールト除去およびフォールト予測は、システムの仕様に示された仕事をいかに忠実に実行することができるか、デペンダビリティの**確証**（dependability validation）を行う方法とみなされる。システムの行う仕事の信頼、またその正当性は、主としてシステムが実行する仕事について、デペンダビリティの**属性** に関するシステムの アセスメント（assessment）を基礎として判定される。

これまでに紹介した概念は、次の３つの主な項目に集約される。（図１）

- **阻害要因**（impairments ）：フォールト，誤り，障害のことでデペンダビリティでない原因あるいは結果をもたらすような望ましくない（期待されない）環境である。（その定義はデペンダビリティの定義から容易に導かれる。仕事に関してはそれ以上正当性（reliance）がないものとみなされる。）
- **手段**（means）：フォールト予防，フォールトトレランス，フォールト除去，フォールト予測による技法・解決方法であって　a)正当性を必要とする仕事を実行する能力を与えること　b)この能力により確実性（confidence）があることが必要である。
- **属性**（attributes）：　信頼度，アベイラビリティ，安全度，セキュリティのことで　a）システムの持つ能力に期待する性質を表し　b）阻害要因による結果及びその対抗手段を定めるシステムの品質である。

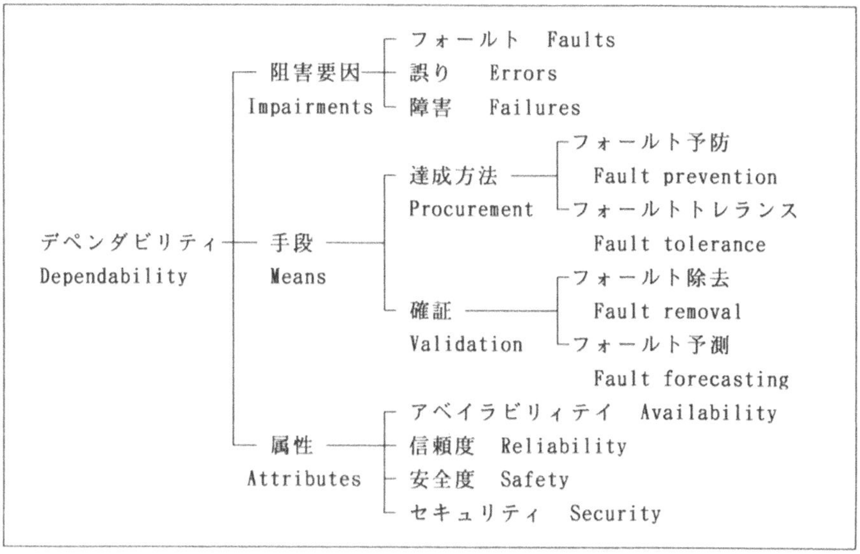

図　1　デペンダビリティの関係図

2．一般概念としてのデペンダビリティについて

あらゆる科学的および技術的な方策に関する一般的な傾向として、まず最初に関連した問題を早急に解決するために、その検討すべき分野を限定するということがある。時が経過するにしたがって他の分野における方策との関連が無視できないようになる。そこで他の分野における方策は、いま検討を行っている方策の特殊なケースにすぎないという大胆な提案が行われる。大いに議論がたたかわされた結果として、それぞれの方策の一貫性を保つために独自の専門用語を生むことになる。このようなことが実際にコンピュータシステムに関する信頼性・安全性およびセキュリティについてもいえる。当初はコンピュータシステムの信頼性が主な関心事であった。コンピュータシステムの信頼性が高くなると、それによる仕事が恒常的に行われる必要性からアベイラビリティが必須条件となった。コンピュータシステムが極めて重要な分野で用いられるようになると安全性が重要視されるようになった。最も安全なシステムは何もしないことであり、それではなんの意義も見出せない。安全性について関心を寄せている人々は、信頼性があれば安全性をみたし得ると考えている。分散システムが出現したことで、セキュリティの面ではむしろ厄介な問題を抱えることになっている。セキュリティを必要とするシステムでは、その機能を果たす必要はいうまでもなく、セキュリティが損なわれることは破局的な影響をもたらすことから、これに関心のある人々は安全性と信頼性があればセキュリティが満たされ得ると考えている。しかしながら実際には信頼性・安全性・セキュリティの関係は、単純なデペンダビリティでは扱えずに複雑な様相を示すことがある。いま仮にいわゆる「ソフト爆弾」なるものを例として取り上げてみよう。コンピュータシステムにフォールトが生じるのは「テロリスト」が定めた時期であり、彼等のなすが儘にシステムの障害をもたらすこととなる。そのシステム障害も利用者がその事実に気が付かなければ大したことはないものと見過ごすこともあろう。この例では明らかに信頼性・安全性・セキュリティを含む複雑な問題であり、着目している視点によってその局面が変化する[1]。
確かなものとして言えることは、利用者はこのようなシステムが実行する仕事は信用できないし、また信用してはならないことであり、これをさらに明確な言葉で表すとデペンダビリティがないといえる。

これまでの議論のもとで明らかにしておきたいことは、信頼性が安全性より広い概念であるとか、あるいはこれにセキュリティを加えれば信頼性より広い意味を持つというような論述を必ずしも目的としているわけではないことである。信頼性・安全性・セキュリティとデペンダビリティとのかかわり合いの本質が何であるかを示すため

---

[1] これについてはＡＣＭソフトウエア工学ノート(ACM Software Engineering Notes)の信頼性・安全性・セキュリティに関する「コンピュータシステムの公共利用に対する危機 (Risk to the Public in Computer Systems) 」の項に報告されている。

のものであって、これまでの前者の轍はこれに続く後者の道を与えるものとみなしたい。できれば、両者にとって有益な成果をもたらすものであってほしい。これがすでに何度かにわたって示した信頼性・アベイラビリティ・安全性・セキュリティなどの用語に関し、さらに説明を加えた主な理由である。また、デペンダビリティを全般的な概念として導入したもう一つの理由は、信頼性という用語が語源的にも信頼できる能力ということから、あまりにも広範囲にわたる意味を持つようになっているため、一般に理解されている信頼性の高いシステム仕事の継続性を表わす用語（歴史的な観点から）として、また確率的な定義に基ずくシステム信頼度という意味に限定することも意図[2] している。

デペンダビリティは信頼性と同義語のようであるが、これは依存性（dependence）に基づく概念を有する。信頼性によって表わされる肯定的な意味に比べるとデペンダビリティは否定的な概念と思われるかもしれないが、現在の社会では、一般的に高度なシステム、とりわけコンピュータシステムに依存する度合いが大きくなっている。語源的にさらに検討を加えてみると"to rely" はフランス語の"relier"から由来するもので、これはラテン語の"religare"の出典による。つまり背に結び付けたものを" 再び（re）" ゆわえつける、結ぶ（to fasten, to tie ）" という意味がある。信頼性のフランス語は "fiabilité " であって、その語源は１２世紀にまでさかのぼることができ、その意味は" 頼りがいのある性質" であってラテン語の"fidare"による。すなわちごく一般的な" 信頼できること（to trust）" の意味になる。以上のような語源的な検討を行ってみると、現在多くの分野において採用されている信頼性という用語の定義は、" 信用（trust ）" の概念を" 能力（ability ）" という概念に置き換えていることは[3] 、次に述べる理由によって好ましくないものと考えられる。

a) システム利用者の見地からすると利用者が実際に関心あることは、現実に実行される仕事などその機能を果たす能力 ではない。

b) システムの製造者の見地からは、設計の中にフォールトが存在する可能性があり、これを許容したいがために、ソフトウェア技術用語に採用されている（IEE 82参照）ことは認めるとしても、「能力」という用語の内容は明確にしない。

---

[2] さらに次のことを記しておくこととする。

a) タイトルとして「信頼性」という用語を用いた図書は、システムの信頼性をいかに評価・測定・予測するかということを扱っており、実際にいかに信頼性の高いシステムを構成するかということに関しては扱っていない。

b) 信頼性・アベイラビリティなどの一般化概念として、またこれらの用語を含むようなデペンダビリティの用語は過去にすでにその試みがある［Hos 60］。しかしながらその試みはここに述べる内容より一般化されていないために、アベイラビリティ・信頼性を包含するような測度にとどまりセキュリティは考慮されていない。

[3] 例えば《信頼性：与えられた条件で与えられた時間間隔に要求された機能を実行するアイテムの能力》［IEC 85］がある。

3. システムの機能・動作・構造及び仕様について

これまでのところ**システム**は全体的なもの、外部からその動作が認識できるものと考えている。このブラックボックス的な見方による定義では、他の事物、即ち他のシステムと相互に動作して相互に入出力を交換するものとみなすことができる。これら他のシステムは、いま考えているシステムの**環境**（environment ）を形成している[4]。システムの**利用者**（user）は、考えているシステムと相互に動作する環境の一部であって、システムに入力を与え、出力を受け取り、システムによって実行した仕事を利用(use the service ）する特性を有するものと解釈される。

システムの**機能**（function）は、システムが実行しようとしている働き［Kui 85］である。システムの**構造**は、実行する内容に従って構成する［Zie 76］。文献［And 81］の趣旨に従うと、構造的な見地（グラスボックス）によるシステムは、動作を行なうための**構成要素**（component ）の集合体であり、その各構成要素は、また別のシステムとみなすことができる。このような見方は再帰的に適用できるが、システムの内部構造についてそれ以上区分できないか、あるいはその必要性がない最小のシステムを**原始的**（atomic）という。構成要素という用語は広義の意味として考えるべきものである。構成要素内部の階層もシステムの階層と同じように扱い、構成要素はそれ自身システムであって、それを構成し相互関係にある構成要素から形成されている。

これまでの古典的なシステム構造の定義はシステムがいかなるものかを示すことである。このような定義では、デペンダビリティをいかにして達成するかということを明らかにする必要がないようなシステムを表わす場合には適しており、構造は固定的なものとして考えられる。しかしながらシステムの構造は固定的なものと限定して扱うことは好ましくない。デペンダビリティの達成のために、あるいはそれを達成したことによって構造が変化することも許容することが必要となる。

---

[4] a）再帰的な定義を与えることは、なにも再帰性を目的としたものではない。見方によっては相互的であることを強調したいためのものである。従ってシステムの概念としては、与えられたシステムの境界は、その設計者・利用者・保守員等の見解によって異なる。

b）システムの環境は時の経過にしたがって、特に製品寿命の各局面によって変化する。例えばプログラミング環境は、システムの物理的環境が動作状態であるときにも与えられた環境として考えても齟齬はない。

訳注　前ページの脚注[3]に関して、信頼性用語 JIS Z 8115 には、

《信頼性：アイテムが与えられた条件で規定の期間中、要求された機能を果たすことができる性質》と与えている。

このことから構造は状態[5]を有するものと思われる。**状態**（state ）の概念に関する定義は、環境の集合体の動作あるいは構造に関する状況といえる[6]。

最初に示した定義（利用者が認識できる仕事）から、システムが実行する仕事は、明らかにシステム動作を抽象化したものである。この抽象化概念は、コンピュータ・システムが適用される応用分野によって大きく変わることは注意しておく必要がある。その一つの例として時間に関する抽象化概念が重要な役割を果たすことがある。システムとその利用者の時間素片（time granularities）に関する認識は一般に差異があり、この差異は、システムの応用分野によって異なる。さらに付け加えると、仕事は単にその出力についてばかりでなく、利用者にとって有用なあらゆる相互作用が含まれる。例えば入力用のスキャンを行うセンサは、モニターシステムが実施する仕事の一部分である。

これまでのところ機能及び仕事について単一なものと考えている。一般にはシステムは複数の機能及び仕事を実行する。機能及び仕事も、より細分化された機能及び仕事の複合体と見なすことができる。このことからここでは簡単のためにシステムは複数の機能及び仕事を実行するものとして扱う。

デペンダビリティのある特別な局面として、時間に関する特性をあげることができる。**リアルタイムでの機能又は仕事**(real-time function or service) をある制限された時間内に環境に対して提供することが要求される。つまり**リアルタイム・システム**(real-time system)はリアルタイムな機能あるいはリアルタイムな仕事を実行するシステムである[PDC90] 。

システムの仕様は、a)期待される機能及び仕事の一方あるいは両方について、システムに期待する内容を記述する。b)さらにそれらに関して満たす必要があるような環境・実施時間・性能・可観測性などについて記述する。機能とか仕事の内容に関する仕様は、いかにその機能を満たし、どの様に仕事を実行するかというように、システムの主要な目的について通常は最初に規定する。システムの安全性とかセキュリティに関して考慮する場合には、どのようなことが 行する必要のある かというような内容（即ち致命的な事態が発生したとき、危険な状態とか曝露してはならないような重要な情報など）を仕様に記述することになる。このような仕様書の内容は、発生してはならないような事態を軽減するために、システムとして満たす必要のある或いは遂発生してはならない付加的な機能とか仕事について規定する（即ち利用者の認可された権利を調べて許可するなど）。

---

[5] a）従って構造は動作(behavier)を伴うものといえる。特にデペンダビリティの阻害要因に関して、利用者の要求側面に対して、阻害要因の側面は技術の革新的進歩により－幸いとも言えるが－ 差異がある。
b）この定義は構造が変化するようなシステム、即ち適応システム（特に学習能力を持つような）にも当てはめることができる。

[6] この定義はいま考えている現象或いは環境に直接依存するような状態の概念との関連性を明白にすることを目的としている。すなわち計算実行中の状態、障害発生時の状態などである。

さらに次のような仕様内容が規定される。

a)様々な段階の記述による表現：要求仕様・設計仕様・実装仕様等.

b)構成要素の障害が生じた場合の対策：正常時における動作の記述、及び障害が生じて生存している資源だけではそれ以上正常な仕事を継続できないような場合の縮退動作等.

このように仕様を定めることにより、一般的にいって仕様書は単一のものではなく複数の仕様書となり、場合によってはそのために仕様の内容が輻輳化して、かえってシステムの障害をもたらすこともあり得る。

仕様は、システムの供給者（ここでは広い意味で、設計者・制作者・販売者など）とシステムの利用者間の二者の個人あるいは法人組織間において合意されたものである[7]。この合意は、実際に行なわれた仕事が受理できるかどうか、或いは同じことではあるが障害が生じていないか生じたかを判定する基準を与えうる仕様であることが必要である。あるレベルで記述された仕様に関して受理可能であっても、よりおおまかに記述されたレベルの仕様には適さないことがある。これは仕様を詳しく記述するときに生じる間違い（mistake），つまり仕様のフォールト（specification fault）によるものである。仕様のフォールトにより、様々な解釈による仕様となる。より一般的にいえば、ひとたび設定した仕様は変更ができないとすることはよくない。製品寿命の間に、何が生じるかあらかじめ予想することは困難であるから、生じた結果をもとに変更することがありうる。システムに対する要求を修正することにより変更が行われる。期待する機能及び仕事，あるいはフォールトの補正によってその修正が行われる[8]。くどいようではあるが仕様は、二者間で合意されることが重要である。

これまでに述べたシステムの構造・機能・仕事・仕様の概念は、そのまま同じように構成要素にまで拡張して適用できる。さらにこのことはハードウェアあるいはソフトウェアにおける設計段階，一般市販（off-the-shelf）構成部品にも延用できる。設計者が着目すべきことは、機能・仕事について、細部（内部）動作よりもむしろこれらによって実施される内容である。

---

[7] この合意はあるシステムを仕様書及び取扱い説明書と共に購入する場合とか既存するシステムを使用する場合にも当てはまるであろう。

[8] ここで再起的な問題に直面する。即ち実行された仕事が受理可能なものかどうかを規定する必要があるが、その規程自身間違っているかもしれない。仕様書を改定することが相当の注意を払って長期にわたって行われること、この目的のために製品の全寿命期間にわたって、これを適用することを明記すべきであろう[Boe 88]。

4. デペンダビリティの阻害要因

4.1. フォールト

フォールト及びその原因はきわめて広範囲にわたる。ここではフォールトについて、その性質（nature），原因（origin），及び存在性（persistence ）の３つに分類する。フォールトの性質は次のように区分できる。

- **偶発フォールト**（accidental fault）偶発的に発生又は生じたフォールト.
- **意図的フォールト**（intentional fault ）恐らくは悪意によって意図的に生じたフォールト.

フォールトの原因は次の３つの見方によって分けることができる。

- 現象的原因（phenomenological causes ）は次のように区分できる［Avi 78］。
  - －**物理的フォールト**（physical faults ）物理的現象として表れるフォールト.
  - －**人為的フォールト**（human-made-faults ）人間の不完全によりもたらされるフォールト.
- システム境界（system boundaries ）を基準として次のように区分できる。
  - －**内部フォールト**（internal faults ）計算処理を実施するとき誤りを発生するシステムの一部の状態.
  - －**外部フォールト**（external faults ）システムのインターフェースから影響を及ぼす物理的な環境（電磁波外乱・放射線・温度・振動等）または人為的な操作によるフォールト.
- システム全体寿命における発生過程（phase of creation ）によって次のように区分できる。
  - －**設計フォールト**(design faults) a)システムの初期設計段階（広い意味では要求仕様に包括すべき条件）及び次の補正の段階、b)システムの動作あるいは保全における手順確立段階のいずれかで不完全であることにより生じるフォールト.
  - －**動作中フォールト**（operational faults）システムの使用中に発生するフォールト.

フォールトの存在性によって次のように分けることができる。

- **固定フォールト**（permanent faults）システムの内部（計算処理中）あるいは外部（環境）の状況にかかわりなく存在するフォールト.
- **一時的フォールト**（temporary faults）システムの内部・外部の状況に関連し、ある一定時間存在するフォールト.

意図的なフォールト（必ずしも意図的と限定しなくともよいが）はセキュリティに関するもので、明らかに人為的なフォールトである。意図的なフォールトは内部あるいは外部によるもので次にその例を示す。

- 内部フォールトに関して意図的な設計フォールトである**故意不良論理**（malicious logic ）、いわゆるトロイの木馬（Trojan horses ）.
- 外部フォールトに関して、意図的な操作外部フォールトの**侵入**（intrusion ）.

意図的なフォールトは、偶発的なフォールトが鍵となって成功することがある。例えば、偶発的な設計フォールトのためにセキュリティの壁を越えて侵入されることがある。これは障壁が欠けていたために、偶発的な一時の外部フォールトとして開張(exploiting)されて侵入されるもので、興味深い例である。判断あるいは推論される

フォールトの分類の定義において、現象的な原因を導入するにあたり、再起的な不毛の長い議論となる。すなわち、「ではなぜプログラマはミスをするのか」とか「集積回路はどうして故障するのか」ということが疑問になる。フォールトの基本的概念は任意に与えられるため、再起的な疑問をどこかで切り分ける必要がある。そこで、フォールトは、誤りの原因であると 誤りの原因を予防ある ものと定義する。この原因は、フォールトトレランスのメカニズム，保守員・修理場・設計者・半導体物理学者等の見方によっても様々であろう。そこで我々の立場では、
いはトレランスにする ものをフォールトとする。この見方は、人為的及び物理的なフォールトを区別するにあたっては矛盾がある。すなわち、コンピュータ・システムは人が作りだすものであり、中に生じるフォールト或いはシステムに影響を与えるものはシステムの動作に関連した現象を全て知りつくすことができず、人の能力が及ばないため、所詮人為的なフォールトとみなせる。厳密な意味からすれば、物理的なフォールトと人為的なフォールト（特に設計フォールト）を区別する必要はないであろう。しかしながら、デペンダビリティを達成し、その効果を上げるための方法及び技法を考える上では、この区別は重要な意味がある。仮に上記の再起的な見方が切り分けられないものとすれば、フォールトは与えられたシステムの仕事に影響したあるいは影響しつつある他のシステム（設計者をも含めた）の障害の結果であるということにする。このような議論では、次のような例をあげることができる。

- 設計フォールトは、設計者による障害の結果である。
- 物理的な内部フォールトは、ハードウェア構成要素の故障によるものである。その故障は、電気的あるいは電子的なレベルでの誤り（希に生じる故障を物理的な信頼性分野では偶発とか予測不可能などのように位置づけている）による結果であり、これは物理化学的な正常でない理由によって生じるものである。さらに細かくみればハードウェアの製造工程あるいは半導体物理に関する知識の限界に基くものといえる。物理的な内部フォールトは再発フォールトと見なすことができる。これは故障した構成要素を同一の故障のない構成要素と交換することにより、修復できるからである。新しい構成要素も設計及び製造段階での障害の原因を明らかにしてこれを取り除かない限り、再び故障となる可能性がある。しかしながら修復した構成要素の故障率は下がっているものと見なすことができる。
- 物理的あるいは人為的外部フォールトは実際には設計フォールトである。これはシステムがその全寿命期間中に直面するあらゆる状況について、予見することができないこと，またある部分については無視（経済的理由から）したことが原因である。
  - 電磁波外乱の場合には、外部フォールトなのか、あるいは設計フォールト（適当なシールドを行なっていない）のか.

－操作員が一文字だけ不適当な文字を入力したために生じた障害の場合は、操作フォールトかあるいは設計フォールト（入力が正しいかどうかシステムから確認しない [Nor 83] ことによるもの）なのか.

一時的に存在するフォールトについては次のような見方がある。

1）物理的な環境から発生する一時的に存在する外部フォールトは、**過渡フォールト**（transient fault ）という。

2）一時的な内部フォールトのことを**間欠フォールト**（intermittent fault）という。このようなフォールトは、希に生じるようなある組み合わせの状態の場合によって引きおこされる。次にその例を示す。

a）半導体記憶素子のパターン依存フォールトは、ハードウェア構成要素のパラメータの変化（温度変化による影響，寄生容量によるタイミングの遅延等）によって生じる。

b）ハードウェアあるいはソフトウェアの影響による場合として、システムの負荷がある一定レベル以上（タイミングとか同期が臨界にあるときなど）となってフォールトが生じる。実際のところこのような場合のフォールトは、フォールト状態を抽象化して表わしたものである。間欠フォールトの基本的概念は、厳密な意味ではむしろ任意に発生するものと見なされる。このようなフォールトは固定フォールト以外のもので、その活性状態は再現性がなく、しかも希にしか生じない。しかしながらすでに物理的フォールトと設計フォールトの区別について述べたように、このことについて検討を行なうことは充分に意義がある。

以上のような議論を基にすれば、いかなるフォールトも固定的な設計フォールトと考えられよう。厳密な意味では正しいかも知れないが、このような言い方をしてもシステムの設計及び評定を行なう人々にとっては、何も役に立たないことになる。

図２は様々な観点からみた考え方による各種フォールトを分類して示したものである。

図２に示した５種類の観点をもとにそれらの組み合わせを考えるとさらに３２の異なったクラスに分類される。実際に意味のある組み合わせとなる図３の各欄に示す１１通りについて、それぞれ見方と通常の意味でのフォールトによって分類される。

４．２．誤り

誤りとは障害をもたらすものとこれまでは定義している。ある誤りは、次に示す三つの大きな要素によって実際に障害をもたらすかどうかが定まる。

1）システムの構成、特に存在する冗長性の性質.

- 意図的な冗長性（フォールトトレランスの導入）により、誤りによって障害を引きおこすことがないような明白な目的がある。
- 意図しない冗長性（冗長性をもたせないでシステム構成を行なった場合には、実際には難しい[9] ）により、期待していないにも拘らず、意図的な冗長性と同じ効果がある。

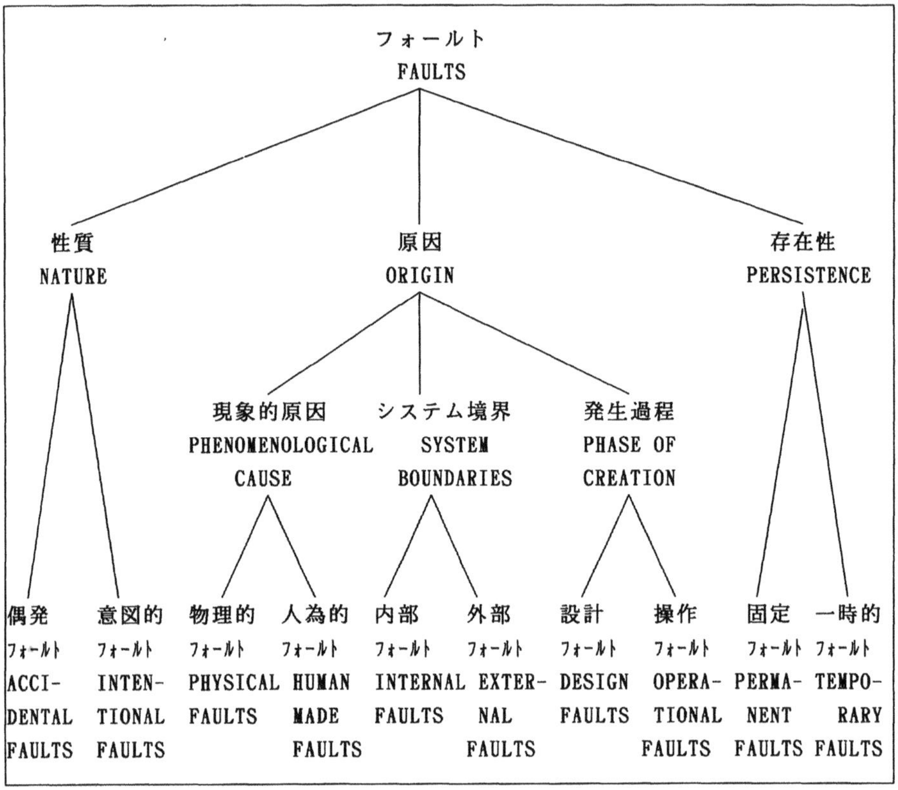

図　2　見方によるフォールトの分類

2）システムの動作により、損害が発生する前に誤りの処理を実施する。

3）利用者観点による障害の定義による、利用者によっては障害は、困惑程度である場合もある。例えば、a)利用者の立場からみた時間単位（time granularity）により、利用者のインターフェースにおいて、誤りは特に異常をもたらさなく通り過ぎる（passes through）ことがあり　b)データ伝送における許容誤り率の概念においては、障害が発生することもあらかじめ考慮してある。これは、システムの利用できない最大時間（利用者の時間単位をもとにして）を仕様の上で明確に規定することで、議論上の立場を明らかにしている。

---

[9] ハードウェアのテストに関する古典的とも言える冗長性を除去する問題がある。これは冗長性のためにフォールトをマスクし、そのためにテスト・パターンの生成が困難になるからである。

性質		原因						存在性		
		現象的原因		システム境界		発生過程				
偶発フォールト	意図的フォールト	物理的フォールト	人為的フォールト	内部フォールト	外部フォールト	設計フォールト	操作フォールト	固定フォールト	一時的フォールト	通常の意味
✓		✓		✓			✓	✓		物理的フォールト
✓		✓			✓		✓	✓		
✓		✓			✓		✓		✓	過渡フォールト
✓		✓		✓			✓		✓	間欠フォールト
✓			✓	✓		✓			✓	
✓			✓	✓		✓		✓		設計フォールト
✓			✓		✓		✓		✓	相互動作フォールト
	✓		✓	✓		✓		✓		故意不良論理
	✓		✓	✓		✓			✓	
	✓		✓		✓		✓	✓		侵入
	✓		✓		✓		✓		✓	

4．3．障害

システム構造の定義にもとずいて、障害をシステムに適用するか或いは構成要素に適用するかという議論は、構成要素それ自身システムであるから、特に意味があるわけではない。原子システムを考えている場合には、基本的な障害と言う概念が最も自然である。

システムは、常に同じように障害を起こすとは限らない。システムが障害となる様相のことを障害モードといい、３種類の見方，領域・システム利用者の認知・環境に及ぼす結果によって性格づけすることができる。

障害領域に関する見方では次のように区分される。

- **値の障害**(value failures)：実行された仕事による値が仕様と一致しない。
- **タイミング障害**(timing failures) ：実行される仕事のタイミングが仕様と合わない。

このような一般的な定義（仕様の内容と適合しない）は**不定障害**(arbitrary failures) にあてはめることができる。例えばタイミング障害は、仕事の実行が速すぎるか遅すぎるかによって早期タイミング障害，遅延タイミング障害のように定義される。

値の障害とタイミング障害の両障害に関して**停止障害**(stopping failure)がある。システムの動作がかりにあったとしても、一定の値の仕事を続けて利用者は何も認知できない。一定の値とは応用に依存するものであって、最後に与えられた値とか或いはあらかじめ定められた値である。停止障害の特殊な場合が**非稼働障害**（omission failure) [Cri 85b,Ezh 86] であり、仕事が全く実行されない。このような障害は、値の障害（０の値）とタイミング障害（無限遅延障害）の両方が生じたものと見なすことができる。永続的な停止障害は**クラッシュ障害**（crash failure ）である。停止障害のみが生じる（あるいはより一般的にそれが許容される）場合のことを**フェイルストップ・システム**(fail-stop system)といい[10]、クラッシュ障害でも許容されるようなシステムのことを**フェイルサイレント・システム**(fail-silent system) [Pow 88]という。

システムに複数の使用者があるときは、障害の認知の仕方により次のように区分される。

- **無矛盾障害**(consistent failures) ：システムの全使用者が同一と認知する障害。
- **不一致障害**(inconsistent failures) ：複数のシステム使用者が同一と認知しない障害。不一致障害［Lam 82］のことを一般にビザンチン障害（Byzantine failures）といっている。

障害が生じた結果により、システムの環境に及ぼす影響を段階ずけすることによって障害の程度を表わし、障害モードの順位づけを行なうことができる。最も一般に受け入れられる次に示す両極端の２種類の故障モードに分類することができる。

- **良性障害**(benign failures) ：障害の結果として、障害のないときにシステムが仕事を実行して得られる利益と同程度の大きさ（一般にはコスト）であるとき。
- **悪性障害**(catastrophic failures) ：障害の結果、障害のないときにシステムが仕事を実行して得られる利益と比較にならない程の大きな損失を被るとき。

障害が良性障害であるときに限る（より一般的には許容できる）システムのことを**フェイルセイフ・システム**(fail-safe system)という。障害程度の概念から、致命的という概念を導くことができる。すなわちシステムの**致命的**(criticality) とはその（可能な）障害モードの最も高度なものである[11]。障害モードと障害程度の関係は応用に深く依存している。しかしながら動作しないときに安全側の位置とするような幅広い応用分野（即ち地上交通とかエネルギ発生工場等）があり、そこではフェイルストップとフェイルセイフとして与えられる技法を用いる[Min 67,Nic 89] 。

---

[10] フェイルストップ・プロセッサの概念は[Sch 83]の分散システムの中で示されている。こここで与えるフェイルストップ・システムの定義は、メッセージ交換により情報を伝送する分散システムを対象としてフェイルストップ・プロセッサの定義と同意なものとする。

[11] 例として航空関係分野では致命的なレベルを次のように定義している[RTC 85]。
- 致命的(critical)：障害が発生した結果、航空機が安全な飛行の継続及び着陸ができない機能。
- 重大(essential) ：障害が発生した結果、航空機の能力低下又は乗務員が操作するにあたり困難さを伴う機能。
- 軽微(non essential) ：障害によっても航空機の性能、乗務員の能力に重大な影響を与えない機能。

4．4．フォールトの因果関係

フォールト，誤り，障害の発生及びその影響の機構は次のように要約することができる。

1）誤りを生じるときはフォールトが**活性化**(active)されている。活性化されたフォールトは、 a)はじめには**不活性な**（dormant ）内部フォールトが計算処理によって活性化（間欠フォールトに関しフォールトの条件が同時に存在する場合を含めて）するか、あるいは、 b)外部フォールトによる内部フォールトが、不活性化及び活性化状態を交互に繰り返すこともある。物理的なフォールトは、ハードウェア構成要素に直接影響を及ぼす。一方人為的フォールトは、あらゆる構成要素に影響が及ぶであろう。

2）誤りは潜在するか検出される。誤りは、誤りであることが認識されるまで**潜在**（latent）し、検出アルゴリズムもしくは検出機構によって**検出**（detect）される。誤りは検出されなければ消滅することもある。一般的には誤りは伝搬し、その伝搬によって他の新たな誤りを生じる。

3）誤りがシステムの使用者のインターフェースに現われる（pass through）場合に障害が発生し、システムの実行している仕事に影響を与える。構成要素の障害は、 a)その構成要素から成るシステムのフォールトであり、また、 b)それと相互に動作する他の構成要素からもフォールトと見なされる。従って障害となっている構成要素の障害モードは、それと相互に関係のある構成要素によって定まるタイプのフォールトとなる。

以上述べたような因果関係は、次に示す鎖状（fundamental chain ）で表現できる。

… → 障害 → フォールト → 誤り → 障害 → フォールト → …

フォールトの因果関係を説明するため、次にいくつかの例を示す。

- プログラマの誤りは（活性化されていない） ソフトウェアの中のフォールト（間違った命令やデータ）として生じている。それが表面化（activation）することによって誤りが影響を与えるようになる（このフォールトの表面化はフォールトのある命令や命令の系列あるいはデータの一部を活性化するような入力パターンによって生じる）。この影響を与えるフォールトが誤ったデータが実行している仕事に影響（仕事による値またはタイミングについて）を与える時、障害が発生する。
- 集積回路に生じた短絡は障害（回路の仕様内容による）である。その結果、ある一定値となったり、関数機能が変更されてそれが表面化しない限り潜在したフォールトである。その後に起こる経過については前例と同じである。
- 大きなエネルギーの電磁波外乱はフォールトである。 a)電磁波外乱により、配線の中に電荷が生じて直接に誤りを生じる。 b)別の種類の（内部）フォールトが生じる。例えば、この外乱がメモリの入力に瞬間的に起きると、メモリが書き込み状態にあればいくつかのビットの値が変り、メモリの中でフォールトとして残る。このフォールトはそのメモリの番地を読み出すまで潜在的に存在する。

外部からの間欠フォールトによる誤りの結果として、電気的レベルの内部フォールトに変化する。

- システム動作中における操作員の不適切なマンマシーン・インターフェイス操作はフォールト（システム側から見て）であり、その結果変更された処理データは誤りである。
- 保守、操作マニュアルの誤りは当該マニュアルのフォールト（内容記述のフォールト）である。マニュアルに存在する誤った内容が潜在的に存在し、マニュアル通りに行なおうとするときに誤りを起こす。

以上のいくつかの例から、潜在誤りの存在期間は故障の内容とかシステムの利用法その他の要因によって大幅に異なることがわかる。

人為的なフォールトは、偶発的かあるいは意図的である。プログラマの誤りについての例をあげて、その起こり得る結果について述べてみよう。悪意のあるプログラマによって仕掛けられた論理的な爆弾は、それが活性化（あらかじめ定められたデータ等）するまで潜在する。その誤りによって記憶容量のオーバーフローとか、プログラムの実行速度の低下を招き、実行すべき仕事はその実行を拒否（denial-of-service ）されて、一種の障害となる。

以上の例は単純化して示したもので、実際にはより複雑な形態となる。次にその例を示す。

a）ある構成要素の中に生じる一つのフォールトは様々な異なった原因がある。物理的な構成要素の固定フォールト（接地するような固定故障など）について調べてみれば次のような例があげられよう。
  - 物理的な障害（スレシュホールドの変化など）
  - 設計フォールトによる誤り（間違ったマイクロ命令）によって、信号が層間を通り抜け、その結果スレシュホールドの変化と同じ影響を与えて短絡回路が生じて二つの回路の出力が短絡する。

b）ある分類に属するフォールトは、その影響により、別の分類に属するフォールトとなることがある。前例において、マイクロ命令を実行する間に生じる誤りの原因であるフォールトは間欠フォールトとなることもある。

c）ある見方において、伝搬する過程ではあまり重要視されない（一時的にすぎない）ことがある。例えば、あるソフトウェアの構成要素を実行している場合、入力誤りを起こす外部フォールトを処理しているとき（例外入力領域内でそれらを処置している場合［Cri 80］）、物理的あるいは人為的なフォールトは重要な意味を持つことはない。

d）ある障害は、いくつかのフォールトの組み合わせによる活性化によって引き起こされる。特にセキュリティに関する事項についてこのことがいえる。コンピュータ・システムに抜け道の扉（アクセス制御をよけて通るなどの方法）が偶然にあるいは意図的に組み込まれている場合は設計フォールトであり、このフォールトは、システムに悪意を持った人間が侵入するまでは潜在することになる。意図的なフォールトとして侵入的なログインが試行され、侵入者がログインに成功すれば、誤りが発生する。

例えばあるファイルを変更（統合性を破壊）し、このファイルを利用する許可を与えられた者が使用しようとするときにその影響が現われて障害となる。

フォールト，誤り，障害の用語、あるいは意味づけについて、次の２項目について説明を加えておくこととする。

a）本文中では、むしろ特殊な形式で用いていると思われるこれらの用語は、ある種の阻害要因について簡潔にしかも明確に表現する場合にも適している。バグ（bug ）[12]、欠陥（defect）、機能低下（deficiency）などのフォールト、中途停止（breakdown)、性能異常（malfunction ）、実行拒否（denial-of-service) などの障害に適用できる。

b）フォールト、誤り、障害という用語を選定した一つの理由は、i)フォールト予防、フォールトトレランス、フォールト診断　ii）誤り検出　iii ）障害率、などのような表現がよく使われるようになっていることに基いている。

最後にこの章で与えた定義は統語論的(syntactic) に与えたことを特に記しておくこととする。従って種々の分類に関する規準に重点を置いたものであって、分類された内容よりむしろ見方の方が重要である。

5．デペンダビリティの手段

5．1．デペンダビリティの意味する依存性

第１章で与えた基本的な定義の中で示した「いかにして（how to´s) 」デペンダビリティを向上させるかということについては未解決のままである。あらゆる活動は人間によるものであるから、所詮は不完全であるものと取り扱っている。この不完全ゆえに依存性（dependency）という考え方に至る。この依存性という表現は、すでに知られている各手法を縦横に駆使して（設計及び実装の各段階において実施されることが望ましい）、デペンダビリティのあるコンピュータシステムを構成することを意味している。この依存性はさらに次のように素描することができる。設計手法及び構成手段によって、いかにフォールトの予防を計っても、（作業内容は完全なものではないので）フォールトは発生する。フォールトの除去も不完全であるから、一般市販のハードウェアまたはソフトウェア構成要素であってももとより不完全であり、フォールト予測が重要である。コンピュータシステムのデペンダビリティの向上のためにはフォールトトレランスを必要とする。更に構成方式を基にしたフォールト除去，フォールト予測などによりデペンダビリティの向上をはかる。しかしながら実際にはこの

---

[12] バグという言葉の特殊なものとして[Gra 86]に示されているように間欠ソフトウエア・フォールト（Heisenbugs、ハイゼンベルグの不確定定理による）を一般的な手法により容易に検出できる固定ソフトウエア・フォールト(Bohrbugs 、ボーア原子のモデルによる）から区別するような場合も含まれる。

ように単純化することができない程に複雑であることはいうまでもない。現在のコンピュータシステムは極めて複雑化しているので、設計及び実装にあたって費用対効果のすぐれた道具だて（広い意味では、許容時間の範囲内で完成する能力をも含む）を必要としている。この道具だてはそれ自体もデペンダビリティでなければならない。

上述したことから、フォールト除去及びフォールト予測の間には密接な関連性があることがわかり、これより**確証**（validation）という用語が与えられる。確証はフォールト除去及びフォールト除去を実施した後の検証（verification）すなわち「VとV」［Boe 79］の意味に限られている。このような場合「システムを正しく作成している」（検証に関して）と「正しいシステムを作成している」（確証に関して）という差異がある[13]。この概念を単純に延用して本文で与えようとする用語を導くことができる。「いま私は正しいシステムを作成しているか？」（フォールト除去）という問に対して「どの位の間正しく動作するか？」（フォールト予測）と逆に問いかけることになろう[14]。フォールト除去は、フォールト予防と密接な関係があり両者をあわせて**フォールト回避**（fault avoidance ）という用語があり、これはフォールトのない（fault free）システムをいかに構成するかを目的とする。フォールトトレランスの手順及び構成について確証を行なう必要性を明らかにするにあたり、フォールト除去及びフォールト予測を同一の機能を持つものとして位置づけることができる。確証という用語は、これらの概念を包括する理解し易いものであって、上述した再起的な重要な問題も確証を確証するといえる。すなわち信用（confidence）し得るシステムを作成するために使用される手法及び道具もいかに信用し得るかということが表わされる。ここで引用する**カバレッジ**（coverage）とは、システムの動作寿命期間中において、実際に直面する状況に対してシステムが有効に動作（保証されている期間）している状況を表わす測度ということにする[15]。不完全なカバレッジという意味は、フォールト除去とフォールト予測の両方に関係がある。フォールト除去のカバレッジが不完全であることからフォールト予測が必要というように考えることができる。

この節に引き続いてフォールトトレランス，フォールト除去及びフォールト予測について検討を加える。フォールト予防については、一般のシステムエンジニアにとって明かなものとしてここでは取り扱わないこととする。

---

[13] 通信のプロトコルの分野などでは［Rud 85］この考え方が逆になっていることを指摘しておきたい。

[14] 有効性(validity)にもとずく確証とは次の二つの概念を要約したものである。
- 与えられた時点での有効性とは、フォールト除去に関係がある。
- 与えられた期間の有効性とは、フォールト予測に関係がある。

[15] ここで定義するカバレッジとはごく一般的なものであり、応用分野にもとずいて正確に表現することができる。
- ソフトウェアテストでのカバレッジとはその内容及び制御フローに関係する。
- 半導体回路素子テストのカバレッジとはフォールトモデルに関係する。
- フォールトトレランスのカバレッジとはフォールトの分類に関係する。
- 設計構想のカバレッジとは実現性と関係する。

5．2． フォールトトレランス

フォールトトレランス[Avi 67]は、誤り処理及びフォールト処置［And 81］によって実現される。**誤り処理**(error processing)はできる限り障害となる前の状態に、計算実行状態のまま誤り除去することを目的とし、**フォールト処置**(fault treatment)は、フォールトが再び活性化することのないように防止することを目的とする。

誤り処理は次に示す二つの手法により実施される。

- **誤り回復**（error recovery）：誤り状態から誤りの無い状態にするためのもので、このような状態復帰には次の２つの方式［And 81］がある。
  - **遡及回復**（backward recovery ）：誤りの生じた状態を誤りの発生する以前にさかのぼって復旧する。そのためには、**回復点**（recovery point）を設定する。これは実行処理を行なう間に、回復のために現在実行しつつある状態を時間的に記録する。
  - **新規回復**（forward recovery）：誤りの生じた状態から新たな状態を見出してシステムは動作することができる（多くの場合は機能低下した動作状態）。
- **誤り補償**（error compensation）：誤り状態でも充分な冗長性があるために誤りの無い場合と同様に仕事を実行することができる。

誤り回復を行なう場合には、誤り状態が変化しない内に検出（できる限り速やかに）する必要がある。この目的のために**誤り検出**（error detection ）があり、誤り検出及び回復の手段が一般に用いられる。構成要素についてこのような処理能力を有することを目的としたハードウェア［Car 68,Wak 78,Nic 89］あるいはソフトウェア［Yau 75,Lap 90a］に関する**セルフチェッキング構成要素**（self-checking component ）がある。セルフチェッキング構成要素がきわめて有用である理由の一つとして、誤り波及範囲（error confinement area）［Sie 82］が明らかに定義できることにある。システム構成において、セルフチェッキング構成要素からなる誤り補償が行なわれている場合、同一のタスクを実施することのできるシステム（いわゆる即時冗長系）がある。このシステムの状態推移は、故障した構成要素を故障のない構成要素に切り換える方法を用いる。このような手法にもとずくフォールトトレランスのことを誤り検出・補償[16]と見なすことができる。一方、誤りが存在しなくとも組織的に補償を行なう手法のことを**フォールトマスキング**（fault masking ）（多数決冗長系など）という。しかしながらこのような手法では、冗長性が知らない内に低下することがある。実際には、マスキングを行なう手法でも誤り検出を行なう手法を用いることが望ましい。もとよりこの検出は状態が推移した後に実施してもよい。

遡及回復と新規回復は両立し得るものである。誤りが発生したならば、まず第一に遡及回復を行ない、次いで新規回復を試行する。新規回復にあたっては、その損害程度を査定する必要がある。一般的にいって遡及回復の場合には、誤った状態から誤りの無い状態への推移は影響のないような機構とするので、損害の査定は無視すること

---

[16] 誤り検出・補償は誤り検出・回復の限定された場合と見ることができよう。ここでの回復は誤りのある状態を単に誤りのない状態に置き換えることをせずに、現在の（誤っている）システムの状態を用いて実施される。

ができる[And 81]。誤り処理を行なうために必要な実行時間に関するオーバーヘッドは、適用する誤り処理の手法によって大幅に変化する。

- 誤り回復において、時間的なオーバーヘッドは誤りが発生することにより長くなる。遡及回復では、誤り処理のためにあらかじめ準備されていた回復点に関係がある。
- 誤り補償について、補償を行なうために必要な時間的オーバーヘッドは、誤りがある場合とない場合と同じか、あるいはほぼ同じ程度である[17]。

さらに誤り補償を実施するに要する時間は、誤り回復に要する時間よりもはるかに短時間でよい。これは冗長性（構成上の）の量が大きいことによる。

以上のことから次の注釈が得られる。

a）実用上重要なこととして、適用すべきフォールトトレランスの方策の選定条件は利用者の時間的単位の設定に関係する。

b）実施上の時間的オーバーヘッドと冗長構成にはある関係がある。より一般的にいえば、冗長システムは実施上の時間的なオーバーヘッドを小さくするため常時冗長な動作を行なっている。利用者から見れば、何等認められない程に時間的オーバーヘッドが小さい。従って実行している仕事は時間的にはなにも冗長がない。上と逆の方式として「時間冗長」（繰り返し動作による冗長性）がある。この場合の構成上の必要な冗長性は初期設定のみでよい。構成上の冗長性を大きくすれば、時間上の冗長性は少なくともよいということがいえよう。

フォールトの処置の第一段階は、**フォールト診断**（fault diagnosis ）である。このフォールト診断は誤りの原因を決定することにあり、その位置の特定とその性質を明らかにすることを表わす用語である。次の段階として、フォールト処置の主要な目的でもあるフォールトの再発生を防止することがある。フォールトを不活性化することが目的であることから**フォールトの不活性化**（fault passivation ）ともいう。これは、さらに影響することがないように明らかにフォールトである構成要素を取り除くことである。システムが以前と同様の仕事を実行することができない場合は、再構成（reconfiguration ）を実施する。

誤り処理によってフォールトを取り除くか、あるいは再発生する可能性が低いと考えられる場合には、フォールトの不活性化を行なう必要はない。フォールトの不活性化が行なわれていない場合に発生するフォールトのことを**ソフトフォールト**（soft fault）といい、不活性化が行なわれている場合のフォールトを**ハード**（hard）あるいは**一定**（solid ）フォールトという。このソフトフォールトとハードフォールトの概念は前に述べた一時的フォールト及び固定フォールトと同義語と思われるであろう。実際のところ、一時的フォールトについてはフォールトトレラント化するために処置をする必要はない。誤り処理は、この場合フォールトの効果を直接取り除くことができて、固定フォールトが生じるような影響を及ぼすことはない。ソフトフォールト及びハードフォールトの概念は次に述べるような理由から有用である。

---

[17] いずれの場合でもシステムの状態を表すレコードを更新するための時間的なオーバーヘッドは加わる。

- 一時的フォールトと固定フォールトを区別することは困難であり、また複雑な手順を要する。 a)一時的なフォールトはフォールト診断を行なう前に、ある一定時間後には消滅する。 b)異なった種類のフォールトでも同じような誤りを引き起こすことがある。このような困難さがあるので、ソフトフォールトはフォールト診断が成功しない場合のフォールトとみなすことができる。
- ソフトフォールトとハードフォールトの概念は、ある種の過渡フォールトの動作モードと微妙な関係にある。例えば、アルファ粒子（回路パッケージの残留イオンによる）あるいは空間中の重イオンにより影響を受けた記憶素子上（拡大解釈してフリップフロップを含む）の不活性内部フォールトは一時的フォールトとみなしてよいかという疑問がある。このような不活性フォールトはソフトフォールトである。

物理フォールトについて前に示したフォールトは、設計フォールトについても適用可能である。設計過程におけるフォールトを想定して、これ等のフォールトトレランスを実際に扱うことができる。フォールトの発生とその活性化に関して冗長化を行なうことで、フォールトの関連性をなくすことができる。

物理的フォールトと設計フォールトに対してフォールトトレランスにする例をあげる。広く用いられているフォールトトレランスの一つの方法として、多数のチャンネルを用いた並列計算処理がある。物理的なフォールトが予測される場合には、ハードウェアの故障は、互いに独立に発生するものとの仮定に基づき、同一の多数チャンネルを用いる。一般的には設計フォールトに対してフォールトトレランスではないことから、異なった設計及び実装 [Elm 72, Ran 75, Avi 78] 、すなわち**ダイバシティ設計** (design diversity) [Avi 84] により同一の仕事を多数のチャンネルで行う方法がある。

フォールトトレラント・システムを扱うに当たって、複数のフォールト或いは故障に遭遇することが多い。このような場合はその原因を明らかにすることを考える。

- **独立複数フォールト**(independent faults)は異なった原因による。
- **関連複数フォールト**(related faults)は共通の原因による。

共通な原因による例としては、電源装置・クロック・仕様書などがある。ダイバシティ設計における関連複数フォールトは、各個別設計及び細部記述にあたり独立性が保持されないことが原因であろう。関連複数フォールトは一般的に共通モード障害 (common-mode failure) の原因となる。

フォールトトレラント・システムについて議論を進めるにあたり興味深い概念の一つとして、**一致複数誤り**(coincident errors) すなわち同一入力によって現れる複数の誤りがある[Avi 86]。

複数の障害を時間的な関係で次のように区別する。

- **同時複数障害**(simultaneous failures) は、ある定められた時間窓の範囲に発生する。
- **逐次複数障害**(sequential failures) は、ある定められた時間窓の範囲外で発生する。

数学的な定義にもとずく同時ということは時間窓がゼロのことである。

しかしながら実際には応用分野によって時間窓を広くとる必要があることから[Tri 84]、ほとんど同時の複数障害として扱う。代表的な例としてある時点の単一フォールトに対応できるフォールトトレラント・システムでは、次のフォールトが発生する以前に前のフォールトについて回復処置をする必要がある。このことから逐次複数障害が同時とならないよう区別するため、時間窓は誤り処理及び可能なフォールト処置を行うため（この期間はシステムは脆弱である）に必要な時間間隔とする。

複数の構成要素を用いて協同動作を行う場合の一つの重要な側面として、誤りの伝搬及び障害のない構成要素の動作に影響を与えることがないように防止することであろう。与えられた構成要素が、個々の目的で他の構成要素と情報の交換を行なうような局面で特に重要な意味がある。単一情報源の例として、ローカルなセンサによるデータ，ローカルなクロック，他の構成要素の状況に関するローカルな見方などがある。単一情報源から発生する情報をある一つの構成要素から複数の構成要素に対して通信を行なう場合に、故障のない構成要素は、互いに矛盾の生じることがないようにして、どのように情報を交換し、それらの合意(agreement)を行なうかということがある。分散化システムの分野では、この種の問題（例えばアトミック・ブロードキャスト[Cri 85a]、クロックの同期問題［Lam 85,Kop 87］、会員制プロトコル［Cri 88］）が重要視されている。しかしながら、いかなるフォールトトレラント・システムにおいても、必然的に冗長構成されているために、様々なレベルの分散化が行なわれるので、合意問題が存在する。地域的に広範囲にわたるフォールトトレラント・システムでは、この合意問題を解決する手段として、古典的な分散システムにおけるメッセージ交換（すなわち中継［Lal 86］，相互合意のための多段中継［Fri 82］）を用いるが、費用がかかりすぎると考えられている。

システムのある特徴を解明することで必要な冗長性の量をおさえることができる。このことを低費用フォールトトレランスという。この特徴の例として構造上の統一性をもとにした誤り検出・訂正符号［Pet 72］、強健データ構造［Tay 80］、マルチプロセッサ及びコンピュータネットワーク［Pra 86,Ren 86］、アルゴリズムに基づくフォールトトレランス［Hua 82］などがある。フォールトトレラント化されるフォールトの対象は、想定したフォールトに直接関係し、生じる特性によって定まる。

利用者にとって構成要素の障害による警告信号は重要である。これについては例外(exception）［Mel 77,Cri 80,And 81］の枠組的な貢献の中で取り扱っている。ある言語体系において例外処理機能をもたせることにより、誤り回復、特に新規回復を実施するために都合がよいことが示されている[18]。

フォールトトレランスも再起的な概念である。本質的にいってフォールトトレランスの目的とする機構は、影響を与えるようなフォールトに対して保護するためのもの

---

[18] 『例外』という用語を用いるにあたって、例外的な場合（必ずしも誤りではない）という本来の意味をそのまま用いることはフォールトトレランスの枠組み的作業の上では注意を要する。フォールトトレランスが当然なものとして備わるコンピュータシステムでは、初期の設計段階から考慮されていて、『例外的』な属性はないとすることに矛盾があるようにみなされるからである。

である。その例として多数決（voter）、セルフチェッキングチェッカ［Car 68］、回復プログラム及びデータのための安定な記憶［Lam 81］等がある。

フォールトトレランスは何も偶発フォールトに限ってはいない。古典的な暗号方式に対する侵入をも保護する［Den 82］ある種の誤り検出機構は、意図的及び偶発フォールトの両者に対して（メモリのアクセス保護技法など）有効である。侵入及び物理フォールトの両方ともに対するフォールトトレランス［Fra 86, Rab 89］、及び故意不良論理に対するフォールトトレランス［Jos 88］などが提案されている。

5.3. フォールト除去

フォールトの除去は3つの段階、すなわち検証（verification），診断（diagnosis），訂正（correction）からなる。**検証**は、システムの満足すべき性質に沿っているかどうかいわゆる検証条件（verification condition）を検定して行なう。この検定が正しくない場合には、検証条件が満たされない原因であるフォールトの診断、次いで必要な訂正の2段階を実施する。訂正を行なった後、フォールト除去により望ましくない結果をもたらさないことを検証して、従来の実行処理に戻る。この段階での検証のことを（非）**復帰検証**（(non-)regression verification）という。検証条件は次の2通りがある。

- 一般条件：システムのあるクラス（これにはデッドロックが無いこと、設計及び実装規約の確認（conformance））に適用される条件である。
- 該当システム特有の条件：システム仕様に直接関連した特殊な条件である。

検証の技法は、システムを実際に動作させるか、あるいは動作させないかにより分類することができる。システムを実際に動作させないで実施する検証のことを**静的検証**（static verification）という。
この検証は次のようなものがある。

- システムそれ自身について、a）静的分析（検査あるいは立入検査）［Mye 79］，データフロー解析［Ost 76］，複雑性解析（complexity analysis）［McC 76］，コンパイラ検査等、b）正確性証明（proof-of-correctness）（帰納法的手法［Hoa 69, Cra 87］）。
- システム動作モデル（ペトリネットとか有限状態オートマトン）による動作解析（behavior analysis）［Dia 82］。

システムを動作させながら検証を行なうことを**動的検証**（dynamic verification）という。システムの入力にシンボルを与えて行なう場合のことを**シンボリック実行**（symbolic execution）といい、実際の与える場合のことを**テスト法**（testing）という。

システムにあらゆる可能な入力を与えてテストを行なうのは一般的にいって実用的ではない。テストパターンを決定する方法は、テスト入力を選定する規約（criteria）とテスト生成法の2通りの見方によって分類される。

テスト入力を選定する規約はさらに次のように3通りに分類される。

- テスト法の目的：システムがその（機能の）仕様を満足しているかを検定する**規格テスト法**(conformance testing) がある。ここでフォールトを見出すことを目的としたテスト法のことを**フォールト発見テスト法**(fault-finding testing) という。

- システムのモデル：システムの機能或いは構造によって**機能テスト**(functional testing)及び**構造テスト**(structural testing)に分けることができる。
- フォールトモデルの存在：**フォールトを基にしたテスト法**(fault-based testing)[Mor 90]を実行する。これには特殊なフォールトを明らかにすること（ハードウェア製造過程における固定フォールト（stuck-at-fault） [Rot 67]，マイクロプロセッサの命令セットに影響を与える物理フォールト [Tha 78]，ソフトウェアの設計フォールト [Goo 75,DeM 78,How 87]）がある。フォールトモデルがない場合の規約には仕事を実行するためのシステムの能力（パス活性化（path sensitization) [Rap 85, Nta 88]、ソフトウェアの入力境界値（input boundary value） [Mye 79] に関して定める[19]。

テスト入力は決定的（deterministic ）あるいは確率的（probabilistic ）に生成する。

- **決定的テスト法**（deterministic testing ）：テストパターンはある規約を適用することによって選択してあらかじめ決定する。
- **ランダム・テスト法**（random testing）または**統計的テスト法**（statistical testing)：テストパターンは、入力領域において定められた確率分布に基づいて選定する。入力データの分布と数は、ある規約を適用して決定する [Dav 81, Dur 84] 。

様々な観点を組み合わせることで数多くのテスト法が存在するが、次に簡略化してそのうちの幾つかを示す。

- フォールトの発見・構造・フォールトを基にした構造テスト法をハードウェアに一般的に適用することは、フォールトの発見・構造・フォールトを基にしないテストをソフトウェアに一般的に適用することに相当する。
- ソフトウェアの変異テスト法(mutation testing)[DeM 78]は、フォールトの発見構造・フォールトを基にした決定的テスト法である。

テスト出力を観測して検証条件を満たしているかどうかということに判定を下すことは、神託問題（oracle problem） [Adr 82] として知られている。検証条件を出力全体に適用して調べるか、コンパクト機能により圧縮化された出力を調べるかの方法がある（ハードウェアの物理的フォールトをテストするシステムのシグニチャ手法 [Dav 86]，あるいはソフトウェアの設計フォールトをテストする 部分的神託（partial oracle） [Wey 82]）。物理的フォールトをテストするにあたり、入力を与えてテストを行なうシステムの出力結果を予測するためには、コンパクト化するとしないにかかわらずシミュレーション [Lev 86] 、または基準となるシステム（ゴールドユニット）を用いる。設計フォールトに関しては、仕様が基準となるので、プロトタイプを作成するとか、ダイバシティ設計により同一仕様に基づいて複数のシステムを作成する（バック対バック・テスト法（back-to-back testing） [Bis 88]）。

---

[19] フォールトモデルが定められている可能性は開発段階と深い係わりあいがある。さらに進歩するとよいフォールトモデルが定義される可能性が高くなる。

ある種の検証方式を複数組み合わせて用いることがある。シンボリックを用いる方法は a)テストパターンの決定を行なうために利用 [Adr 82] または b)正確性証明 [Car 78] に用いられる。

検証は、システムの開発と併行して実施されるので、これまで述べた手法は開発段階において様々な形態、プロトタイプ・構成要素等について適用される。システムの検証は、仕様に示されている以外については実施できないが、意図的フォールトについて検証する [Gas 88] ことも特に重要である。

システムの設計にあたり、検証を容易に行なうため、**検証可能設計** (design for verifiability) を行なう。物理的フォールトに対するハードウェアについて開発が行なわれており、対応する手法として **テスト容易化設計** (design for testability) [Wil 83,McC 86] といっている。

システムの全寿命期間中の動作段階におけるフォールト除去のことを訂正保全 (corrective maintenance) といい、仕様に従った仕事を実行する能力の維持及び向上を目的としている[20]。訂正保全は次の二つの方式がある。

- **修治保全** (curative maintenance) ：誤りをもたらしたフォールトを除去し、そのことを報告する。
- **予防保全** (preventive maintenance) ：誤りをもたらす前にフォールトを除去する。このようなフォールトは次のものがある。
  - －前回予防保全を行なった後に発生した物理的フォールト、
  - －他の同じようなシステムで誤りをもたらした設計不良 [Ada 84] 。

これらの定義は[21]、フォールトトレランスでないシステムはもとより、オンラインのまま（実行している仕事を停止させることなく）あるいはオンラインで保全を行なうようなフォールトトレラントシステムにも適用される。

訂正保全とフォールト処置の間に定かに境界を与えることはできない。特に修治保全は、結局のところフォールトトレランスを達成する一手段と考えられよう。

---

[20] 保全に関してはさらに次ぎのような区分がある[Ram 84]。
- 適応保全(adaptive maintenance)：システムを環境の変化に応じて調整する（例えばオペレーテイング・システムとかデータベース・システムを変更）。
- 完全保全(perfective maintenance)：システムの機能を客側（及び設計者）の変更要求に応えて、仕様のフォールトの除去等が含まれる。

[21] 『保全』という用語をソフトウェアに当てはめると、不適当であるように考えられるが単語の持つ語源的意味をとらえればわかりやすい。中年の世代では、保全とは兵器が戦場で使用可能状態を保持することを意味しており、事後・適応・完全保全のことである。実際に（現状からの）逸脱からハードウェアを修理する場合の保全についていえば、仕事の概念に関しそれを「保持」することであるから語源的な意味からも適切であり、特にその語源について議論の余地はない。

5．4． フォールト予測

フォールトの予測は、フォールト発生または活性化によるシステムの動作を評価（evaluation）することによって行なわれる。この評価は次に示す二通りの局面から行なう。

- 非確率的：フォールトトリーのパスの最小カットセットを決定し、障害モード及びその影響について検討する。
- 確率的：デペンダビリティの属性に関する確率的、すなわちデペンダビリティの測度と定義されている要素についてシステムの動作性能を明らかにする。

システムの全寿命期間にわたり、仕様に従って仕事を実行する次の２つの状態の交互に繰り返しと利用者は認識する。

- **正しい仕事**(correct service）：実行している仕事は仕様と一致している[22]。
- **正しくない仕事**(incorrect service）：実行している仕事は仕様と一致していない。

従って、障害は正しい仕事から正しくない仕事を実行する状態への推移であり、その逆は**復旧**（restoration)である。正しい仕事と正しくない仕事の実行の数量的表現は、デペンダビリティの測度と定義されている信頼度及びアベイラビリティである。

- **信頼度**（reliability ）：連続的に正しい仕事を行なう測度であって、障害となるまでの時間とも同意である。
- **アベイラビリティ**（availability）：正しい仕事と正しくない仕事を行なう場合において、正しく仕事を実行する場合の測度である。

第３の測度として一般的に**保全度**（maintainability ）がある。これは障害が発生してから、又は正しくない仕事を連続して実行していることから復旧までに要する時間の測度である。

安全度（safety）に関する測度は信頼度を拡張したものとみなせる。正しくない仕事を行なう状態となっても良性障害となり、安全な状態（悪性損害あるいは危険とはならないという意味で）となるシステムを考える。このとき**安全度**とは連続的に安全となる測度、同じように悪性障害となるまでの時間である。従って安全度とは、悪性障害に関する信頼度と考えることができる。アベイラビリティの拡張として、悪性障害が生じた後の正しくない仕事と安全との交互の状態に関する測度もあり得るが、これは大して意味のない測度である。悪性障害が発生した場合は、その結果は一般に極めて重大であり、仕事の復旧は次の２つの理由から、主要事項とは考えられない。

- 悪性障害の結果を修復することは、次順位（広い意味で法律的な側面まで含めると）にすぎない。
- システムを再動作させるためには（委員会調査等により）長時間かかることもあり、算定できない程の損失となる。

---

[22] ここでは『正しい』という用語はシステムが実行する仕事に限定して用い、システムそのものについては用いない。これは欠陥のないシステムは存在しないと言っても過言ではなく、システムが未だに故障しないにすぎないと見做されるからである。

ここで信頼度とアベイラビリティの混在した測度を定義することができる。良性障害が生じた後、正しい仕事と正しくない仕事について正しい仕事を行なう測度であるとすれば、この測度は悪性障害に至るまでのシステムのアベイラビリティを量的に与えており、いわゆる信頼度（あるいはアベイラビリティ）と安全度のトレードオフを量的に評価することができる。

多重化機能を有するシステムでは、仕事を実行する上でいくつかのモードが認められる。これは最大限に仕事を行なうことから、正しい仕事をわずかしか実行しない範囲がある。性能に関係するデペンダビリティの測度として**パーフォーマビリティ**（performability）という用語が使用されている［Mey 78,Smi 88 ］。

デペンダビリティの測度を量的に評価する事を目的として、確率的なフォールト予測に２通りの手法、モデル化と（評価）テスト法がある。これらの間には互いに相補関係がある。すなわちモデル化のために、基本的なプロセス・モデル（障害のプロセス、保全プロセス、システムの活動化プロセス等）を与え、これを基にしてテスト法により得られたデータが必要である。

モデル化により評価を行うにあたり、対象として考えるシステムが次に示すように安定信頼度であるか、或いは信頼度成長であるかによって大きな差異がある[Lap 90]。

- **安定信頼度**（stable reliability）：正しい仕事を実行するシステムの能力が一定であるもの（障害に至るまでの継続時間が確率的に同一であるもの）。
- **信頼度成長**（reliability growth）：正しい仕事を実行するシステムの能力が改善されるもの（障害に至るまでの継続時間は確率的に増大した値となるもの）[23]。

安定信頼度及び信頼度成長について実用上の立場から解釈をすると次のようになる。

- 安定信頼度：復旧を行なったとき、システムは復旧する以前と同一の能力である。次に示すような状況に相当する。
  －ハードウェア障害の場合、故障した部品は同一の故障のない部品と交換される。
  －ソフトウェア障害の場合、障害となった入力とは異なった入力のパターンでシステムを再実行させる。
- 信頼度成長：障害を引き起こす活性化フォールトについて、設計フォールト（ソフトウェアあるいはハードウェアの）との診断されれば、これを除去する。

安定信頼度に関してシステムのデペンダビリティを評価するにあたって、一般につぎに示す主要な２段階がある。

- システムのモデルを構築：基礎的な確率過程をもとにシステムの構成要素の動作及びその相互動作をモデル化する。

[23] 信頼度の低下（正しい仕事を実行するシステムの能力の低下により、障害に至るまでの継続時間の確率が低下）は理論的なものである。実際的には訂正を行っている間に新たなフォールトが取り込まれて、それが活性化する確率のほうが取り除いたフォールトよりも増大することがある。このような場合でも信頼度の低下は時間的な制限があり、長期間にわたつて見れば信頼度は成長していると見做すことができよう。

・モデルを基に処理：得られた表現からシステムのデペンダビリティの測度の値を求める。

評価は、 a)物理的フォールト［Tri 84］、b)設計フォールト［Lit 79, Arl 88］、c) 物理的フォールトと設計フォールトの両方の組み合わせ［Lap 84, Pig 88］について実施することができる。

システムのデペンダビリティはその環境に大きく依存しており、広い意味での利用期間［Hec 87］、あるいはより特殊な例として負荷［Cas 81, Iye 82］によって変わる。

信頼度成長について多くのモデルがハードウェア［Dua 64］，ソフトウェア及びその両方について［Lap 90］提案されている。その多くはソフトウェアに関するものであって、信頼度［Yam 85,Mil 86 ］あるいは、（残留）フォールト数［Goe 79, Toh 89］に関するものである。これらのモデルは、すでに与えられた過去の障害データを基に将来の信頼度を予測する事を目的としており、予測問題（prediction problem）［Lit 88］に関係あるものとして特に注目されている。

システムの検証が（主な）目的でないとしても、評価テスト法[Mil 87, Cho 87]が5.3 節で定義したような見方によって、即ち規格・機能・フォールトに基かない・静的なテストとして利用することができる。ここで注目に値するものとして、入力を与える側面により、操作上の側面を表すことから**操作テスト法**(operational testing)と呼ばれるものがある。

フォールトトレラントシステムを評価する場合には、誤り処理及びフォールト処置機構のカバレッジも重要な要件［Bou 69,Arm 73 ］であり、その評価はモデル化［Dng 89］あるいは、フォールト挿入（fault injection ）［Arl 89］と呼ばれるテスト法によって行なうことができる。

6. デペンダビリティの属性

属性はすでに第１章で様々な性質として定義している。この性質はいま考えているコンピュータ・システムの応用分野によって、その軽重に差があると思われる。

・アベイラビリティは常に必要であるが、それでも応用によってその程度は変化する。

・信頼度・安全度・セキュリティは応用によって必要・不必要があろう。

他の性質は十分に満たした上で、さらに付け加えておくべき性質として統合性があること、即ち a)データまたはプログラムにおいて、 b)偶発または故意フォールトにより不整合（広い意味で）が生じないことがある。

デペンダビリティの属性を重要視する程度は、システムのデペンダビリティ向上のためにこれまでに述べた各種手法を適切に組み合わせてバランスよく採り入れることに直接的に影響を受ける。これはさらに難しい問題であるが、ある属性は、互いに相反する関係（例えば、アベイラビリティと安全度、アベイラビリティとセキュリティ）があり、その間の費用対効果を考慮する必要がある。

コンピュータシステムの設計にあたっては、費用・性能及びデペンダビリティの3つの重要な次元があるものと考えられるが、費用・性能の次元は当然のこととして、ともすればデペンダビリティに関する次元について関心がうすいことは、問題である[Sie 82]。

保全度については、すでに 5.4節にデペンダビリティの測度の一つとして定義している。保全度は実際に保全を実施する上での作業の行い易さにも関係する性質であると考えられる。

第1章に与えたセキュリティの定義に関し、許可されていない使用、或いは情報の漏洩の防止等について提起[ITS 90]されている。この中でセキュリティを秘密保持、情報・統合性の許可されない漏洩・破壊の防止、情報の許可されない修正・削除の防止、許可されない情報の取得保存の防止等の組み合わせとして定義している。さらにここでつぎのことを指摘しておくこととする。

a）許可されない情報の取り扱い及び使用は、偶発或いは故意のフォールトによって引き起こされることがある。 5.2節でも示したように無許可使用の防止機構は、この両方のフォールトのいずれにも共通的に弱い。

b）意図的フォールトについては、許可という概念は広く理解されているので、許可を与えられた個人がその職権を乱用することは、規則に違反する行為として許可権を実際に行使していることになる。

フォールトシステムの能力に関する重要な特徴として、誤り検出技法が備わっているために、システムが実行した仕事（機能）が正しいかどうかという情報を提供することができることがある。このような性質のことを**信用性**(trustability)[Fur 90]という。

システムが実際にデペンダビリティがあるかどうか評定すること、つまり実行された仕事を正しいものとして信用できるかということは次に示す3つの理由により前章で述べた確証を行なう方法より上位に属する。

- 設計のゆきとどいた範囲、また実情に合う評価手法（例えば、テスト入力の決定にあたって使用される実際のフォールトに関する想定、フォールトトレランス機構の設計におけるフォールトの仮定など）を確実に検定するためには、利用する技術に関しシステムの利用目的に沿っているかなど一般的な知識より以上にすぐれた知識と能力を有する必要がある。
- ある種のフォールトに関してシステムのデペンダビリティの属性を評価することは、現段階では実際に役に立たないとか、意味のある結果は得られないと考えられる。確率論に基づく手法は存在しないし、またその手法は広く受け入れられていない。そのいくつかの例として、偶然の設計フォールトに対しての安全とか、意図的フォールトに対するセキュリティ等があげられる。
- 確証を行なうべき仕様は、一般にいかなるシステムでもフォールトはないとされている。

ここに述べた多くの事実を基に次ぎのような説明ができる。

- システムを査定するにあたり強調すべきことは、開発および製造段階で手法および技術をいかに採用してそれを利用するかということである。

ある場合には a) 採用された手法および技術、b)利用手法の査定したがってシステムに等級をつける[24]。

- ある種のフォールトトレラントシステム仕様については、デペンダビリティの測度の項目に対処可能なフォールトの数の確率を加えることがある[25]。もとより上述した限界が満足される場合は、このような仕様は必要ないであろう。

おわりに

高度に進歩し、しかも高性能なコンピュータシステムが広く個人および企業において用いられるようになり、それらによりもたらされる仕事 －現金支払機・人工衛星軌道計算・航空機の制御・原子力発電所の制御・高度情報の管理等－ に深く依存するようになりつつある。異なった環境では、これらの仕事に関して重点指向する性質も異なる。例をあげると、平均的な実時間応答の必要性、要求した結果が与えられる尤度、システム環境を破局に至らすような障害発生を回避する能力、侵入を防止する能力の程度等があげられる。デペンダビリティの概念は、このような様々な要求を一つの枠組みの中で纏め上げるのに都合の良い手法を与える。デペンダビリティは信頼度・アベイラビリティ・安全度・セキュリティ等の性質を特殊な場合として包含している。さらにまたシステムの利用者が一般にシステムに要求するこれらの性質をバランス良く定める課題に関しても、これを表現する手段を与えている。

---

24 この例として次のものがある。

- システムはセキュリティの観点[Dod 85]より、Ａ１（検証済み設計）からＤ（最小保護に等級づけされている。
- 商用航空機に用いられるソフトウエアは、ソフトウエアとして具備すべき機能の致命的な程度として、致命的・重要・通常にしたがってレベル１・２・３のような等級をつけている[RTC 85]。

25 このような仕様について航空の応用分野では古くから「フェイル・オペレーショナル」とか「フェイル・セイフ」などという言い方をしている。

用　語

注意事項　この用語は文書の理解を容易にするためのものであり、本文書と独立に扱ってはならない。

訳　注　　訳出にあたっては従来の用語および現在検討が進められている用語にできうる限り従っている。特にJIS 信頼性用語と異なっている場合はJIS 用語に＊をつけて付記してある。

英文	和文	意味
Accidental fault	偶発フォールト	偶発的に発生したフォールト
Active fault	活性化フォールト	誤りをもたらすフォールト
Arbitrary failure	不定障害	障害の項参照のこと
Atomic system	原子システム	システムの内部構造についてそれ以上区分できないか或いはその必要性がない最小のシステム
Attributes of dependability	デペンダビリティの属性	デペンダビリティの阻害要因から、その達成手段により、システム品質を維持するための属性 信頼度、アベイラビリティ、保全度、安全度、セキュリティ
Availability	アベイラビリティ	使用可能な正しい状態．正しい状態と正しくない状態において、正しく仕事を行うことのできる比率
Avoidance (fault-)	（故障）回避	フォールトのないシステムとするための手法及び技術．フォールト予防及びフォールト除去
Backward recovery	遡及回復	誤り回復の形態であって、誤った状態から過去に遡って元の状態に戻すこと
Behavior (system-)	（システム）動作	システムがなにか実行していること
Benign failure	良性障害	障害による損害が、正しく仕事を実行した時の費用と同一程度の障害
Catastrophic failure	悪性障害	障害による損害が正しく仕事を実行する時の費用とは比較にならないほど大きな障害
Coincident errors	一致複数誤り	同一入力による複数の誤り
Common-mode failures	共通モード障害	関連フォールトが原因の障害

Compensation (error-)	（誤り）補償	誤りの処理形態で、誤った状態でも正しい仕事を実行する上で十分な情報を持つこと
Component (system-)	（システム）部分	他のシステム
Conformance testing	規格テスト法	システムがその仕様を満たすかどうかを検査する目的のテスト法
Consistent failure	無矛盾障害	全システム使用者が同一と認知する障害
Correct service	正しい仕事	システムの仕様に従って実行する仕事
Corrective maintenance	訂正保全 ＊事後保全	システムの動作寿命期間中に仕様にあった仕事を実行するための保全または改善．システムの動作寿命期間中におけるフォールトの除去
Coverage	カバレッジ	システムの動作寿命期間中において、実際に直面する状況に対してシステムが有効に動作する状況を表す比率
Crash failure	クラッシュ障害	永続的な非動作障害
Criticality (system-)	（システムの）致命	最も高度な障害モード
Curative maintenance	修治保全	誤りが生じ、その報知があったときその誤りを除去する訂正保全
Dependability	デペンダビリティ	与えられた仕事を実行するにあたり正確な信頼のおける資質
Design diversity	ダイバシティ設計	システム製造にあたり、同一の仕事を別々の異なった設計・組込みによって行うこと
Design fault	設計フォールト	人間が原因となっている内部のフォールト
Design for verifiability	検証可能設計	システム設計に当たって、その検証を行う機能を備えた方法及び技法
Detection (error)	（誤り）検出	システムの誤りを認識するための動作
Detected error	検出誤り	検出アルゴリズムまたは機構によって誤りが認められたこと
Deterministic testing	限定テスト法	あらかじめ選定され定められたテストパターンでテストする手法
Diagnosis (fault-)	（フォールト）診断	誤りの原因となった場所及び性質を明らかにするテスト手法
Dormant fault	不活性フォールト	計算処理によっては活性化されない内部フォールト
Dynamic verification	動的検証	システムを動作させながら行う検証

Environment (system-)	(システム)環境	与えられたシステムと相互動作する他のシステム
Error	誤り	障害をもたらす可能性のあるシステムの状態.システムにおいてフォールトが表面化すること
External fault	外部フォールト	環境の妨害・干渉により生じるフォールト
Fail-safe system	フェイルセイフシステム	障害を起こしてもある許容範囲となる良性障害のシステム
Fail-silent system	フェイルサイレントシステム	障害を起こしてもある許容範囲で仕事を何もしないような障害となるシステム
Fail-stop system	フェイルストップシステム	障害を起こしてもある許容範囲で停止する障害のシステム
Failure	障害	システムの仕様で定められた仕事から、行った仕事が逸脱すること.正しい仕事の実施から不正な仕事の実施にいたる経過
Fault	フォールト	誤りの決定的原因或いは仮想的原因. 除去或いは許容する誤り原因. 相互動作する他のシステムの障害による当該システムに与える影響
Fault-based testing	フォールトを基にしたテスト法	特定の種類のフォールトを明らかにする目的のテスト法
Fault-finding testing	フォールト発見テスト法	フォールトを明らかにする目的のテスト法
Forecasting (fault-)	(フォールト)予測	フォールトの現在の数・将来の影響・結果等について予測する手法及び技法
Forward recovery	新規回復	誤り回復の形態であって、誤った状態から新たな状態を見出だして移行すること
Function (system-)	(システム)機能	システムが実行しようとしているもの
Functional testing	機能テスト法	テストの実施形態であって、テストの入力はシステム機能に関する定義をもとにして選定される
Hard fault or solid fault	ハードフォールト一定フォールト	必然的に顕在化するフォールト
Human made fault	人によるフォールト	人間の不完全からもたらされるフォールト

Impairments to dependability	デペンダビリティの阻害要因	デペンダビリティを損なう好ましくないが予測できない原因或いは結果
Incorrect service	正しくない仕事	システムの仕様に従わないで実行される仕事
Inconsistent failure	不一致障害	システム使用者が同一とは認知しない障害
Independent faults	独立複数フォールト	異った原因による複数フォールト
Integrity	統合性	減損の無い状態
Intentional fault	意図的フォールト	意図的に発生または生じたフォールト
Intermittent fault	間欠フォールト	一時的に生じる内部フォールト．活性化状態が再現しないとか希に生じるフォールト
Internal fault	内部フォールト	システム内部のフォールト
Intrusion	侵入	意図的な動作上の外部フォールト
Latent error	潜在誤り	認知されていない誤り
Maintainability	保全度	連続して正しくない仕事を実行する尺度．障害が生起してから復帰するまでの時間の尺度
Malicious logic	故意不良論理	意図的な設計フォールト
Masking (fault-)	（フォールト）マスキング	系統的に誤りを補償して誤りのない状態にすること
Means for dependability	デペンダビリティの達成手段	a)システムが信頼のおける仕事を実行できるようにして、b)この能力について確信できるようにするための手法及び技法．フォールト予防、フォールトトレランス、フォールト除去、フォールト予測
Measures of dependability	デペンダビリティの測度	阻害要因が生じてもこれに対抗する手段によって仕事の品質を保つ事のできる属性．信頼度、アベイラビリティ、保全度、安全度
Omission failure	非稼働障害	仕事を実行しない障害
Operational fault	動作中フォールト	システムの使用中に発生するフォールト
Operational testing	操作テスト法	システムのデペンダビリティを評価するために実行するテスト法であって、動作状態を入力側から表示させる
Passivation (fault-)	（フォールト）の不活性化	フォールトを活性化できないようにするための手段
Performability	パフォーマビリティ	実行に関するデペンダビリティの測度

Permanent fault	固定フォールト	システムの外部或いは内部にかかわりなく存在するフォールト
Physical fault	物理フォールト	物理的な現象として表れるフォールト
Preventive maintenance	予防保全	フォールトが活性化する前にそれらを除去するための修正保全
Prevention (fault-)	（フォールト）予防	フォールトの発生或いは生起を予防するための手法及び技法
Processing (error-)	誤り処理	システムから誤りを取除く処置・手段
Procurement of dependability	デペンダビリティの達成方法	システムの仕様に従う仕事を実行する能力をシステムに持たせる事を目的とした手法および技法
Random testing	ランダム・テスト法	統計的テスト法を参照
Real-time function	実時間機能	外部から見て有限時間間隔内に満たすよう要求される機能
Real-time service	実時間での仕事	外部から見て有限時間間隔内に満たすよう要求される仕事
Real-time system	実時間システム	実時間機能或いは実時間での仕事を実行するシステム
Recovery (error-)	（誤り）回復	誤り状態から誤りの無い状態にするための誤り処理の形態
Recovery point	回復点	処理実行中に現在の状態を回復するため必要とする時間上の点
Regression verification	復帰検証	修正を行った後に望ましくない結果となっていないことの検証
Related faults	関連複数フォールト	共通原因に起因する複数フォールト
Reliability	信頼性、信頼度	連続的に仕事を行う上でのデペンダビリティ．連続的に正しい仕事を行う尺度．障害となるまでの時間の尺度
Reliabilty growth	信頼度成長	正しい仕事を実行できるシステムの能力が改善されること
Removal (fault-)	（フォールト）除去	存在するフォールト（数、程度）を減少させる目的の手法および技法
Restoration (service-)	（仕事）複旧	正しくない仕事から正い仕事への移行
Random testing	ランダム・テスト法	統計テスト法参照
Safety	安全性、安全度	悪性障害が発生することのないようなデペンダビリティ．良性障害が生じた後でも正しい仕事あるいは正しくない仕事を連続的に実行する尺度． 悪性障害となるまでの時間の尺度

Security	セキュリティ	確信性および統合制を保護することに関するデペンダビリティ
Self-checking component	セルフチェッキング構成要素	誤り検出機能を内部にもつ構成要素
Sequential failures	逐次複数障害	ある設定された時間窓の範囲外で生じる複数の障害
Service	仕事、サービス	システム利用者が認識できるシステムの動作
Severity (failure-)	(障害) 程度	システム環境における障害の段階
Simultaneous failures	複数同時障害	ある設定された時間窓の範囲内で生じる複数の障害
Soft fault	ソフトフォールト	不活性化されていないフォールト
Solid fault	一定フォールト	ハードフォールトの項目参照
Specification (system-)	(システム) 仕様	システムの要求について合意した記述
State (system-)	(システム) 状態	環境に関してシステムのあるべき条件
Stable reliability	安定信頼度	正しい仕事を実行する能力をシステムが保持すること(障害に至るまでの継続時間と確率的な意味で同じ)
Static verification	静的検証	システムを動作させないで実施する検証
Statistical testing	統計的テスト法	定めた確率分布に基いて入力に与えるテスト・パターンを選定してテストを実施する手法
Stopping failure	停止障害	システムの動作がもはや利用者に認められず一定の仕事しかしない障害
Structure (system-)	(システム) 構造	実行する内容に従ってシステムを構成すること
Structual testing	構造テスト法	システムの構造に従って系統的にテスト・パターンを選定して行うテスト手法
Symbolic execution	シンボリック実行	シンボリック入力で動的に検証を実施すること
System	システム	他の実体と相互に動作し、或いは相互に作用できる実体. 相互動作するための部品の集合体
Temporary fault	一時的フォールト	ある一定時間存在するようなフォールト
Testing	テスト法	有効な入力で動的に検証する手法

Timing failure	タイミング障害	仕事を実行するにあたりタイミングが仕様とあわなくなるような障害
Tolerance (fault-)	(フォールト)トレランス	フォールトの存在にも拘らず仕様に示された仕事を遂行できるような手法及び技法
Transient fault	過渡フォールト	一時的な物理的外部フォールト
Treatment (fault-)	(フォールト)処置	フォールトが再度活性化しないように保護するための手段
Trustability	信用性	仕事が正しく実行されたことに関する情報を利用者に提供するシステムの能力
Un-dependability	非デペンダビリティ	与えられた仕事を正しく忠実に実行できないようなコンピュータ・システムの性質
User (system-)	(システム)利用者	当該システムと相互に作用する別のシステム(物、人)
Validation	確証(確認)	システムの仕様に示された仕事を忠実に実行することを確かめる手法及び技法
Value failure	値の障害	仕様に一致しない仕事による値となる障害
Verification	検証	a)仕様とは関係のない一般的な、或いは、b)仕様を基にした要求に合う特性(検証条件)にシステムが合致しているかどうかを決定する手順

REFERENCES

Ada 84 E. N. Adams, "Optimizing preventive service of software products", *IBM Journal of Research and Development,* vol. 28, no. 1, Jan. 1984, pp. 2-14.

Adr 82 W.R. Adrion, M.A. Branstad, J.C. Cherniavsky, "Validation, verification, and testing of computer software", *Computing Surveys*, vol. 14, no. 2, June 1982, pp. 159-192.

And 81 T. Anderson, P.A. Lee, *Fault Tolerance — Principles and Practice,* Prentice Hall, 1981.

Arl 88 J. Arlat, K. Kanoun, J.C. Laprie, "Dependability evaluation of software fault-tolerance", in *Proc. 18th IEEE Int. Symp. on Fault Tolerant Computing (FTCS-18),* Tokyo, June 1988, pp. 142-147; also in *IEEE Trans. on Computers*, vol. 39, no. 4, April 1990, pp. 504-513.

Arl 89 J. Arlat, Y. Crouzet, J.C. Laprie, "Fault injection for dependability validation of fault-tolerant computing systems", in *Proc. 19th IEEE Int. Symp. on Fault Tolerant Computing (FTCS-19),* Chicago, June 1989, pp. 348-355.

Arn 73 T.F. Arnold, "The concept of coverage and its effect on the reliability model of repairable systems", *IEEE Trans. on Computers,* vol. C-22, June 1973, pp. 251-254.

Avi 67 A. Avizienis, "Design of fault-tolerant computers", in *Proc. Fall Joint Computer Conf.*, 1967, pp. 733-743.

Avi 78 A. Avizienis, "Fault tolerance, the survival attribute of digital systems", *Proceedings of the IEEE,* vol. 66, no. 10, Oct. 1978, pp. 1109-1125.

Avi 84 A. Avizienis, J.P.J. Kelly, "Fault tolerance by design diversity: concepts and experiments", *Computer,* vol. 17, no. 8, Aug. 1984, pp. 67-80.

Avi 86 A. Avizienis, J.C. Laprie, "Dependable computing: from concepts to design diversity", *Proceedings of the IEEE,* vol. 74, no. 5, May 1986, pp. 629-638.

Bis 88 P.G. Bishop, "The PODS diversity experiment", in *Software Diversity in Computerized Control Systems*, U. Voges editor, Wien: Springer-Verlag, 1988, pp. 51-84.

Boe 79 B.W. Boehm, "Guidelines for verifying and validating software requirements and design specifications", in *Proc. EURO IFIP'79,* London, Sep. 1979, pp. 711-719.

Boe 88 B.W. Boehm, "A spiral model of software development and enhancement", *IEEE Computer*, May 1988, pp. 61-72.

Bou 69 W.G. Bouricius, W.C. Carter, P.R. Schneider, "Reliability Modeling Techniques for Self-Repairing Computer Systems", in *Proc. 24th ACM National Conf., 1969*, pp. 295-309.

Car 68 W.C. Carter, P.R. Schneider, "Design of dynamically checked computers", in *Proc. IFIP'68 Cong.*, Amsterdam, 1968, pp. 878-883.

Car 78 W.C. Carter, W.H. Joyner, D. Brand, H.A. Ellozy, J.L. Wolf, "An improved system to verify assembled programs", in *Proc. 8th IEEE Int. Symp. on Fault Tolerant Computing (FTCS-8)*, Toulouse, France, June 1978, pp. 165-170.

Car 79 W.C. Carter, "Fault detection and recovery algorithms for fault-tolerant systems", in *Proc. EURO IFIP'79*, London, Sep. 1979, pp. 725-734.

Car 82 W.C. Carter, "A time for reflection", in *Proc. 12th IEEE Int. Symp. on Fault Tolerant Computing (FTCS-12)*, Santa Monica, California, June 1982, p. 41.

Cas 81 X. Castillo, D.P. Siewiorek, "Workload, performance, and reliability of digital computing systems", in *Proc. 11th IEEE Int. Symp. on Fault Tolerant Computing (FTCS-11)*, Portland, Maine, June 1981, pp. 84-89.

Che 81 M.H. Cheheyl, M. Gasser, G.A. Huff, J.K. Miller, "Verifying security", *Computing Surveys*, vol. 13, no. 3, Sep. 1981, pp. 279-339.

Cho 87 C.K. Cho, *Quality programming*, Wiley, New York, 1987.

Cra 87 D. Craigen, "Strengths and weaknesses of program verification systems", in *Proc. 1st European Software Engineering Conf.*, Strasbourg, France, Sep. 1987, pp. 421-429.

Cri 80 F. Cristian, "Exception handling and software fault tolerance", in *Proc. 10th IEEE Int. Symp. on Fault Tolerant Computing (FTCS-10)*, Kyoto, Japan, Oct. 1980, pp. 97-103; also in *IEEE Trans. on Computers*, vol. C-31, no. 6, June 1982, pp. 531-540.

Cri 85a F. Cristian, H. Aghili, R. Strong, D. Dolev, "Atomic broadcast: from simple message diffusion to Byzantine agreement", in *Proc. 15th IEEE Int. Symp. on Fault Tolerant Computing (FTCS-15)*, Ann Arbor, Michigan, June 1985, pp. 200-206.

Cri 85b F. Cristian, "Exceptions, failures and errors", *Technique et Science Informatiques*, vol. 4, no. 3, 1985, pp. 385-390; in French, English version avail. as IBM Reseach Report no. RJ 4130, Sep. 1983.

Cri 88 F. Cristian, "Agreeing on who is present and who is absent in a synchronous distributed system", in *Proc. 18th IEEE Int. Symp. on Fault Tolerant Computing (FTCS-18)*, Tokyo, June 1988, pp. 206-211.

Dav 81 R. David, P. Thévenod-Fosse, "Random testing of integrated circuits", *IEEE Trans. on Instrumentation and Measurement*, vol. IM-30, no. 1, March 1981, pp. 20-25.

Dav 86 R. David, "Signature analysis for multiple output circuits", *IEEE Trans. on Computers*, vol. C-35, no. 9, Sep. 1986, pp. 830-837.

DeM 78 R.A. DeMillo, R.J. Lipton, F.G. Sayward, "Hints on test data selection: help for the practicing programmer", *Computer*, April 1978, pp. 34-41.

Den 82 D. E. Denning, *Cryptography and Data Security*, Addison-Wesley, 1982.

Dia 82 M. Diaz, "Modeling and analysis of communication and cooperation protocols using Petri net based models", *Computer Networks*, vol. 6, no. 6, Dec. 1982, pp. 419-441.

DoD 85 Department of Defense Standard, "Department of defense trusted computer system evaluation criteria", DOD 5200.28-STD, Dec. 1985.

Dua 64 J.T. Duane, "Learning curve approach to reliability monitoring", *IEEE Trans. on Aerospace*, vol. 2, 1964, pp. 563-566.

Dug 89 J.B. Dugan, K.S. Trivedi, "Coverage modeling for dependability analysis of fault-tolerant systems", *IEEE Trans. on Computers*, vol. 38, no. 6, June 1989, pp. 775-787.

Dur 84 J.W. Duran, S.C. Ntafos, "An evaluation of random testing", *IEEE Trans. on Software Engineering*, vol. SE-10, no. 4, July 1984, pp. 438, 444.

Elm 72 W.R. Elmendorf, "Fault-tolerant progtamming", in *Proc. 2nd IEEE Int. Symp. on Fault Tolerant Computing (FTCS-2)*, Newton, Massachusetts, June 1972, pp. 79-83.

Ezh 86 P.D. Ezhilchelvan, S.K. Shrivastava, " A characterisation of faults in systems", in *Proc. 5th Symp. on Reliability in Distributed Software and Database Systems*, Los Angeles, Jan. 1986, pp. 215-222.

Fra 86 J.M. Fray, Y. Deswarte, D. Powell, "Intrusion tolerance using fine-grain fragmentation-scattering", in *Proc. 1986 IEEE Symp. on Security and Privacy*, Oakland, April 1986, pp. 194-201.

Fri 82 S.G. Frison, J.H. Wensley, "Interactive consistency and its impact on the design of TMR systems", in *Proc. 12th IEEE Int. Symp. on Fault Tolerant Computing (FTCS-12),* Santa Monica, California, June 1982, pp. 228-233.

FTC 82 Special session "Fundamental concepts of fault tolerance", in *Proc. 12th IEEE Int. Symp. on Fault Tolerant Computing (FTCS-12),* Santa Monica, California, June 1982:
D.E. Morgan, "Report of subcommittee on models, fundamental concepts, and terminology", pp. 3-5.
A. Avizienis, "the four-universe information system model for the study of fault tolerance", pp. 6-13.
H. Kopetz, "The failure fault model", pp. 14-17.
J.C. Laprie, A. Costes, "Dependability: a unifying concept for reliable computing", pp. 18-21.
A.S. Robinson, "A user oriented perspective of fault tolerant system models and terminologies", pp. 22-28.
T. Anderson, P.A. Lee, "Fault tolerance terminology proposals", pp. 29-33.
P.A. Lee, D.E. Morgan, editors, "Fundamental concepts of fault tolerant computing", pp. 34-38.

Fur 90 D.A. Fura, A.K. Somani, "Trustability: a dependability measure for systems with failure reporting capability", University of Washington, Seattle, Tech. Rep. EE-FTL-90-03, 1990.

Gas 88 M. Gasser, *Building a Secure Computer System,* Van Nostrand Reinhold, 1988.

Goe 79 A.L. Goel, K. Okumoto, "Time-dependent error-detection rate model for software and other performance measures", *IEEE Trans. on Reliability,* vol. R-28, no. 3, Aug. 1979, pp. 465-484.

Gol 82 J. Goldberg, "A time for integration", in *Proc. 12th IEEE Int. Symp. on Fault Tolerant Computing (FTCS-12),* Santa Monica, California, June 1982, p. 42.

Goo 75 J.B. Goodenough, S.L. Gerhart, "Toward a theory of test data selection", *IEEE Trans. on Software Engineering,* vol. SE-1, no. 2, June 1975, pp. 156-173.

Gra 86 J.N. Gray, "Why do computers stop and what can be done about it?", in *Proc. 5th Symp. on Reliability in Distributed Software and Database Systems,* Los Angeles, Jan. 1986, pp. 3-12.

Gun 89 U. Gunneflo, J. Karlsson, J. Torin, "Evaluation of error detection schemes using fault injection by heavy-ion radiation", in *Proc. 19th IEEE Int. Symp. on Fault Tolerant Computing (FTCS-19),* Chicago, June 1989, pp. 340-347.

Hec 87 H. Hecht, E. Fiorentino, "Reliability assessment of spacecraft electronics", in *Proc. 1987 Annual Reliability and Maintainability Symp.*

Hoa 69 C.A.R. Hoare, "An axiomatic basis for computer programming", *Communications of the ACM*, vol. 12, no. 10, Oct. 1969, pp. 576-583.

Hos 60 J.E. Hosford, "Measures of dependability", *Operations Research,* vol. 8, no. 1, 1960, pp. 204-206.

How 87 W.E. Howden, *Functional Program Testing and Ananlysis,* McGraw-Hill, 1987.

Hua 82 K.K. Huang, J.A. Abraham, "Low cost schemes for fault tolerance in matrix operations with processor arrays", in *Proc. 12th IEEE Int. Symp. on Fault Tolerant Computing (FTCS-12),* Santa Monica, California, June 1982, pp. 330-337.

IEC 85 IEC 191, *Reliability, Maintainability and Quality of Service, chapter 191 of the International Electrotechnical Vocabulary,* Document 1-IEV-191-Central Office-1243 and 56-IEV-191-Central Office-119, Geneva: International Electrotechnical Commission, 1985.

IEE 82 IEEE Std. 729, *IEEE Standard Glossary of Software Engineering Terminology,* New York: IEEE, 1982.

ITS 90 *Information Technology Security Evaluation Criteria,* Harmonized criteria of France, Germany, the Netherlands, the United Kingdom, May 1990.

Iye 82 R.K. Iyer, S.E. Butner, E.J. McCluskey, "A statistical failure/load relationship: results of a multi-computer study", *IEEE Trans. on Computers,* vol. C-31, July 1982, pp. 697-706.

Jes 77 D.C. Jessep, "Fault-tolerant computing, definition of terms", Report IEEE Computer Society P610/DI, Feb. 1977.

Jos 88 M.K. Joseph, A. Avizienis, "A fault tolerance approach to computer viruses", in *Proc. 1988 Symp. on Security and Privacy,* Oakland, April 1988, pp. 52-58.

Kop 87 H. Kopetz, W. Ochsenreiter, "Clock synchronization in distributed real-time systems, *IEEE Trans. on Computers,* vol. C-36, no. 8, Aug. 1987, pp. 933-940.

Kui 85 B. Kuipers, "Commonsense reasoning about causality: deriving behavior from structure", in *Qualitative Reasoning about Physical Systems,* D.G. Bobrow editor, MIT Press, 1985, pp. 169-203.

Lal 86 J.H. Lala, "A byzantine resilient fault tolerant computer for nuclear power plant applications", in *Proc. 16th IEEE Int. Symp. on Fault Tolerant Computing (FTCS-16),* Vienna, Austria, July 1986, pp. 338-343.

Lam 81 B.W. Lampson, "Atomic transactions", in *Distributed Systems — Architecture and implementation,* Lecture Notes in Computer Science 105, Berlin: Springer-Verlag, 1981, chap. 11.

Lam 82 L. Lamport, R. Shostak, M. Pease, "The Byzantine generals problem", *ACM Trans.on Programming Languages and Systems,* vol. 4, no. 3, July 1982, pp. 382-401.

Lam 85 L. Lamport, P.M. Melliar-Smith, "Synchronizing clocks in the presence of faults", *Journal of the ACM,* vol. 32, no.1, Jan. 1985, pp. 52-78.

Lap 84 J.C. Laprie, "Dependability evaluation of software systems in operation", *IEEE Transactions on Software Engineering,* vol. SE-10, no. 6, Nov. 1984, pp. 701-714.

Lap 85 J.C. Laprie, "Dependable computing and fault tolerance: concepts and terminology", in *Proc. 15th IEEE Int. Symp. on Fault Tolerant Computing (FTCS-15),* Ann Arbor, Michigan, June 1985, pp. 2-11.

Lap 89 J.C. Laprie, "Dependability: A Unifying Concept for Reliable Computing and Fault Tolerance", in *Dependability of Resilient Computing Systems*, Blackwell Scientific Publications, T. Anderson editor, 1989, pp. 1-28.

Lap 90a J.C. Laprie, J. Arlat, C. Beounes, K. Kanoun, "Definition and analysis of hardware- and software-fault-tolerant architectures", *IEEE Computer,* vol. 23, no. 7, July 1990, pp. 39- 51.

Lap 90b J.C. Laprie, C. Beounes, M. Kaaniche, K. Kanoun, "The transformation approach to the modeling and evaluation of the reliability and availability growth of systems in operation", in *Proc. 20h IEEE Int. Symp. on Fault Tolerant Computing (FTCS-20),* Newcastle, UK, July 1990; extended version in *IEEE Trans. on Software Engineering,* "The KAT (knowledge-action-transformation) approach to the modeling and evaluation of reliability and availability growth", vol. 17, no. 4, April 1991, pp. 370-382.

Lev 86 Y. Levendel, "Fault simulation", in *Fault-Tolerant Computing, Theory and Techniques*, D.K. Pradhan editor, Englewood Cliffs: Prentice Hall, 1986, pp. 184-264.

Lit 79 B. Littlewood, "Software reliability model for modular program structure", *IEEE Trans. on Reliability,* vol. R-28, Aug. 1979, pp. 241-246.

Lit 88 B. Littlewood, "Forecasting software reliability", in *Software Reliability Modeling and Identification,* S. Bittanti editor, Springer-Verlag, 1988, pp. 140-209.

McC 76 T.J. McCabe, "A complexity measure", *IEEE Trans. on Software Engineering,* vol. SE-2, no. 4, Dec. 1976, pp. 308-320.

McC 86 E.J. McCluskey, "Design for testability", in *Fault-Tolerant Computing, Theory and Techniques*, D.K. Pradhan editor, Englewood Cliffs: Prentice Hall, 1986, pp. 95-183.

Mel 77 P.M. Melliar-Smith, B. Randell, "Software reliability: the role of programmed exception handling", *ACM SIGPLAN Notices*, vol. 12, no. 3, March 1977, pp. 95-100; also in *Reliable Computer Systems,* S.K. Shrivastava editor, Springer-Verlag, 1985, pp. 143-153.

Mey 78 J.F. Meyer, "On evaluating the performability of degradable computing systems", in *Proc. 8th IEEE Int. Symp. on Fault Tolerant Computing (FTCS-8),* Toulouse, France, June 1978, pp. 44-49.

Mil 87 H.D. Mills, M. Dyer, R.C. Linger, "Cleanroom software engineering", IEEE Software, Sep. 1987, pp. 19-25.

Mil 86 D.R. Miller, "Exponential order statistic models of software reliability growth", *IEEE Trans. on Software Engineering,* vol. SE-12, no. 1, Jan. 1986, pp. 12-24.

Min 67 H. Mine, Y. Koga, "Basic properties and a construction method for fail-safe logical systems", *IEEE Trans. on Electron. Computers,* vol. EC-16, no. 6, June 1967, pp. 282-289.

Mor 90 L.J. Morell, "A theory of fault-based testing", *IEEE Trans. on Software Engineering*, vol. 16, no. 8, Aug. 1990, pp. 844-857.

Mye 79 G.J. Myers, *The Art of Software Testing*, John Wiley & Sons, 1979

Nic 89 M. Nicolaidis, S. Noraz, B. Courtois, "A generalized theory of fail-safe systems", in *Proc. 19th IEEE Int. Symp. on Fault Tolerant Computing (FTCS-19),* Chicago, USA, June 1989, pp. 398-406.

Nor 83 D.A. Norman, "Design rules based on analyses of human error", *Communications of the ACM*, vol. 26, no. 4, April 1983, pp. 254-258.

Nta 88 S.C. Ntafos, "A comparison of some structural testing strategies", *IEEE Trans. on Software Engineering,* vol. SE-14, no. 6, June 1988, pp. 868-874.

Ost 76 L.J. Osterweil, L.D. Fodsick, "DAVE — A validation error detection and documentation system for Fortran programs", *Software Practice and Experience,* Oct.-Dec. 1976, pp. 473-486.

PDC 90 PDCS Workshop Report no. W6, "Real-time systems, specific closed workshop", Sept. 1990, vol. 1, ESPRIT Basic Research Action no. 1092.

Pet 72 W.W. Peterson, E.J. Weldon, *Error-Correcting Codes,* MIT Press, 1972.

Pig 88 P.I. Pignal, "An analysis of hardware and software availability exemplified on the IBM 3725 communication controler", *IBM J. of Research and Development,* vol. 32, no. 2, March 1988, pp. 268-278.

Pow 88 D. Powell, G. Bonn, D. Seaton, P. Verissimo, F. Waeselynck, "The Delta-4 approach to dependability in open distributed computing systems", in *Proc. 18th IEEE Int. Symp. on Fault Tolerant Computing (FTCS-18),* Tokyo, Japan, June 1988, pp. 246-251.

Pra 86 D.K. Pradhan, "Fault-tolerant multiprocessor and VLSI-based system communication architectures", in *Fault-Tolerant Computing, Theory and Techniques,* D.K. Pradhan editor, Englewood Cliffs: Prentice Hall, 1986, pp. 467-576.

Rab 89 M.O. Rabin, "Efficient dispersal of information for security, load balancing and fault tolerance", *Jounal of the ACM,* vol. 36, no. 2, April 1989, pp. 335-348.

Ram 84 C.V. Ramamoorthy, A. Prakash, W.T. Tsai, Y. Usuda, "Software engineering: problems and perspectives", *IEEE Computer,* Oct. 1984, pp. 191-209.

Ran 75 B. Randell, "System structure for software fault tolerance", *IEEE Trans. on Software Engineering,* vol. SE-1, no. 2, June 1975, pp. 220-232.

Ran 78 B. Randell, P.A. Lee, P.C. Treleaven, "Reliability issues in computer system design", *Computing Surveys,* vol. 10, no. 2, June 1978, pp. 123-165.

Ran 86 B. Randell, "Reliability and security issues in distributed computing systems", in *Proc. 5th Symp. on Reliability in Distributed Software and Database Systems,* Los Angeles, Jan. 1986, pp. 113-118.

Rap 85 S. Rapps, E.J. Weyuker, "Selecting software test data using data flow information", *IEEE Trans. on Software Engineering*, vol. SE-11, no. 4, April 1985, pp. 367-375.

Ren 86 D. A. Rennels, "On implementing fault-tolerance in binary hypercubes", in *Proc. 16th IEEE Int. Symp. on Fault Tolerant Computing (FTCS-16)*, Vienna, Austria, July 1986, pp. 344-349.

RTC 85 *Software considerations in airborne systems and equipment certification*, Document no. RTCA/DO-178A, Radio Technical Commission for Aeronautics, March 1985.

Rot 67 J.P. Roth, W.G. Bourricius, P.R. Schneider, "Programmed algorithms to compute tests to detect and distinguish between failures in logic circuits", *IEEE Trans. on Electronic Computers*, vol. EC-16, Oct. 1967, pp. 567-579.

Rud 85 H. Rudin, "An informal overview of formal protocol specification", *IEEE Communications Magazine*, vol. 23, no. 3, March 1985, pp. 46-52.

Sch 83 R.D. Schlichting, F.B. Schneider, "Fail-stop processors: an approach to designing fault-tolerant computing systems", *ACM Trans. on Computing Systems*, vol. 1, no. 3, Aug. 1983, pp. 222-238.

Sie 82 D.P. Siewiorek, D. Johnson, "A design methodology for high reliability systems: the Intel 432", in D.P. Siewiorek, R.S. Swarz, *The Theory and Practice of Reliable System Design*, Digital Press, 1982, pp. 621-636.

Smi 88 R.M. Smith, K.S. Trivedi, A.V. Ramesh, "Performability analysis: measures, an algorithm, and a case study", *IEEE Trans. on Computers*, vol. 37, no. 4, April 1988, pp. 406-417.

Tay 80 D.J. Taylor, D.E. Morgan, J.P. Black, "Redundancy in data structures: improving software fault-tolerance", *IEEE Trans. on Software Engineering*, vol. SE-6, no. 6, Nov. 1980, pp. 383-394.

Tha 78 S.M. Thatte, J.A. Abraham, "A methodology for functional level testing of microprocessors", in *Proc. 8th IEEE Int. Symp. on Fault Tolerant Computing (FTCS-8)*, Toulouse, France, June 1978, pp. 90-95.

Toh 89 Y. Tohma, K. Tokunaga, S. Nagase, Y. Murata, "Structural approach to the estimation of the number of residual faults based on the hyper-geometric distribution", *IEEE Trans. on Software Engineering*, vol. SE-15, no. 3, March 1989, pp. 345-355.

Tri 84 K.S. Trivedi, "Reliability evaluation for fault-tolerant systems", in *Mathematical Computer Performance and Reliability,* G. Iazeolla, P.J. Courtois, A. Hordijk, Eds. Amsterdam: North Holland, 1984, pp. 403-414.

Wak 78 J.F. Wakerly, *Error-Detecing Codes, Self-Checking Circuits, and Applications*, New York: Elsevier North-Holland, 1978

Wey 82 E.J. Weyuker, "On testing non-testable programs", *The Computer Journal*, vol. 25, no. 4, 1982, pp. 465-470.

Wil 83 T.W. Williams, "Design for testability — A survey", *Proceedings of IEEE*, vol. 71, no. 1, Jan. 1983, pp. 98-112.

Yam 85 S. Yamada, S. Osaki, "Software reliability growth modeling: models and applications", *IEEE Trans. on Software Engineering,* vol. SE-11, no. 12, Dec. 1985, pp. 1431-1437.

Yau 75 S.S. Yau, R.C. Cheung, "Design of self-checking software", in *Proc. 1975 Int. Conf. on Reliable Software*, Los Angeles, USA, April 1975, pp. 450-457.

Zie 76 B.P. Ziegler, *Theory of modeling and simulation,* New York: John Wiley, 1976.

MULTI-LINGUAL CROSS INDEX

English	French	German
Accidental fault	Faute accidentelle	Zufällige Fehlerursache
Active fault	Faute active	Aktive Fehlerursache
Arbitrary failure	Défaillance arbitraire	Willkürlicher Ausfall
Atomic system	Système atomique	Atomares System
Attributes of dependability	Attributs de la sûreté de fonctionnement	Kenngrößen der Zuverlässigkeit
Availability	Disponibilité	Verfügbarkeit
Avoidance (fault ~)	Evitement des fautes	Vermeidung (Fehlerursachen~, Fehler~)
Backward recovery	Reprise	Rückwärtsfehlerbehebung
Behavior (system ~)	Comportement (d'un système)	Verhalten (System~)
Benign failure	Défaillance bénigne	Unkritischer Ausfall
Catastrophic failure	Défaillance catastrophique	Kritischer Ausfall
Coincident errors	Erreurs coïncidentes	Gleichzeitige Fehler
Common-mode failures	Défaillances de mode commun	Common-mode-Ausfälle
Compensation (error ~)	Compensation (d'erreur)	Kompensation (Fehler~)
Component (system ~)	Composant (d'un système)	Komponente (System~)
Conformance testing	Test de conformité	Konformitätstest
Consistent failure	Défaillance cohérente	Konsistenter Ausfall
Correct service	Service correct	Korrekte Leistung
Corrective maintenance	Maintenance corrective	Instandsetzung
Coverage	Couverture	überdeckung
Crash failure	Défaillance par écrasement	Total-Ausfall
Criticality (system ~)	Criticité (d'un système)	Kritikalität (System~)
Curative maintenance	Maintenance curative	Heilende Instandsetzung
Dependability	Sûreté de fonctionnement	Zuverlässigkeit
Design diversity	Diversification fonctionnelle	Entwurfsdiversität
Design fault	Faute de conception	Entwurfsfehler
Design for verifiability	Conception en vue de la vérification	Verifikationsgünstiger Entwurf
Detection (error ~)	Détection (d'erreur)	Erkennung (Fehler~)
Detected error	Erreur détectée	Erkannter Fehler
Deterministic testing	Test déterministe	Deterministisches Testen
Diagnosis (fault ~)	Diagnostic (de faute)	Diagnose (Fehler~)
Dormant fault	Faute dormante	Inaktiver Fehler (inaktive Fehlerursache)
Dynamic verification	Vérification dynamique	Dynamische Verifikation
Environment (system ~)	Environnement (d'un système)	Umgebung (System~)
Error	Erreur	Fehler / Fehlzustand
External fault	Faute externe	Externe Fehlerursache
Fail-safe system	Système sûr en présence de défaillance	Ausfallsicheres System, Fail-safe-System
Fail-silent system	Système silencieux sur défaillance	Fail-silent-System
Fail-stop system	Système à arrêt sur défaillance	Fail-stop-System

Italian	Japanese
Guasto accidentale	偶発フォールト
Guasto attivo	活性化フォールト
Fallimento arbitrario	不定障害
Sistema atomico	原子システム
Attributi della garanzia di funzionamento	デペンダビリティの属性
Disponibilita'	アベイラビリティ
Modo di evitare il guasto	（故障）回避
Recupero indietro	遡及回復
Comportamento del sistema	（システム）動作
Fallimento benigno	良性障害
Fallimento catastrofico	悪性障害
Errori coincidenti	一致複数誤り
Fallimenti a modo comune	共通モード障害
Compensazione dell'errore	（誤り）補償
Componente del sistema	（システム）部分
Test di conformita'	規格テスト法
Fallimento consistente	無矛盾障害
Servizio corretto	正しい仕事
Manutenzione correttiva	訂正保全、＊事後保全
Copertura	カバレッジ
Fallimento per omissione persistente	クラッシュ障害
Criticita' del sistema	（システムの）致命
Manutenzione curativa	修治保全
Garanzia di funzionamento	デペンダビリティ
Diversit' di progetto	ダイバシティ設計
Guasto di progetto	設計フォールト
Progetto per verificabilita'	検証可能設計
Rilevazione dell'errore	（誤り）検出
Errore rilevato	検出誤り
Test deterministico	限定テスト法
Diagnosi di guasto	（フォールト）診断
Guasto inattivo	不活性フォールト
Verifica dinamica	動的検証
Ambiente del sistema	（システム）環境
Errore	誤り
Guasto esterno	外部フォールト
Sistema a fallimento non pericoloso	フェイルセイフ・システム
Sistema a fallimento tacito	フェイルサイレント・システム
Sistema a fallimento con blocco	フェイルストップ・システム

English	French	German
Failure	Défaillance	Ausfall / Versagen
Fault	Faute	Fehler / Fehlerursache
Fault-based testing	Test basé sur des fautes	Fehlerbasiertes Testen
Fault-finding testing	Test pour trouver des fautes	Fehlererkennendes Testen
Forecasting (fault ~)	Prévision des fautes	Vorhersage (Fehler~)
Forward recovery	Poursuite	Vorwärtsfehlerbehebung
Function (system ~)	Fonction (d'un système)	Funktion (System~)
Functional testing	Test fonctionnel	Funktionelles Testen
Hard fault or solid fault	Faute dure ou solide	Harter Fehler
Human-made fault	Faute humaine	Fehlerursache aufgrund menschlicher Einflüsse / menschliche Fehler
Impairments to dependability	Entraves à la sûreté de fonctionnement	Beeinträchtigung der Zuverlässigkeit
Incorrect service	Service incorrect	Fehlerhafte Leistung
Inconsistent failure	Défaillance incohérente	Inkonsistenter Ausfall
Independent faults	Fautes indépendantes	Unabhängige Fehler(ursachen)
Integrity	Intégrité	Integrität
Intentional fault	Faute intentionnelle	Absichtlicher Fehler
Intermittent fault	Faute intermittente	Intermittierender Fehler
Internal fault	Faute interne	Interner Fehler
Intrusion	Intrusion	Eindringen / Einbruch
Latent error	Erreur latente	Latenter Fehler
Maintainability	Maintenabilité	Instandhaltbarkeit
Malicious logic	Logique maligne	Bösartig fehlerhafte Logik
Masking (fault ~)	Masquage (de faute)	Maskierung (Fehler~)
Means for dependability	Moyens pour la sûreté de fonctionnement	Zuverlässigkeitsmittel
Omission failure	Défaillance par omission	Auslassungsausfall
Operational fault	Faute opérationnelle	Benutzungsfehler(ursache)
Operational testing	Test opérationnel	Nutzungstest
Passivation (fault ~)	Passivation (de faute)	Ausgliederung (Fehlerursachen~)
Performability	Performabilité	Leistungsfähigkeit
Permanent fault	Faute permanente	Permanenter Fehler
Physical fault	Faute physique	Physikalische Fehlerursache
Preventive maintenance	Maintenance préventive	Vorbeugende Instandsetzung
Prevention (fault ~)	Prévention (des fautes)	Verhinderung (Fehler~)
Processing (error ~)	Traitement (d'erreur)	Behandlung (Fehler~)
Procurement of dependability	Fourniture de la sûreté de fonctionnement	Zuverlässigkeitsverfahren
Random testing	Test aléatoire	Statistisches Testen
Real-time function	Fonction temps réel	Echtzeit-Funktion
Real-time service	Service temps réel	Echtzeit-Leistung
Real-time system	Système temps réel	Echtzeit-System
Recovery (error ~)	Recouvrement (d'erreur)	Behebung (Fehler~)

Italian	**Japanese**
Fallimento	障害
Guasto	フォールト
Test basato sul guasto	フォールトを基にしたテスト法
Test di identificazione del guasto	フォールト発見テスト法
Previsione del guasto	（フォールト）予測
Recupero in avanti	新規回復
Funzione del sistema	（システム）機能
Test funzionale	機能テスト法
Guasto forte	ハード・フォールト
Guasto causato dall'uomo	人によるフォールト
Impedimenti alla garanzia di funzionamento	デペンダビリティの阻害要因
Servizio non corretto	正しくない仕事
Fallimento inconsistente	不一致障害
Guasti indipendenti	独立複数フォールト
Integrita'	統合性
Guasto intenzionale	意図的フォールト
Guasto intermittente	間欠フォールト
Guasto interno	内部フォールト
Intrusione	侵入
Errore latente	潜在誤り
Manutenzionabilita'	保全度
Logica maliziosa	故意不良論理
Mascheramento del guasto	（フォールト）マスキング
Mezzi per la garanzia di funzionamento	デペンダビリティの達成手段
Fallimento per omissione	非稼働障害
Guasto operativo	動作中フォールト
Test operazionale	操作テスト法
Disattivazione del guasto	（フォールト）の不活性化
Prestazione-affidabilita' congiunte	パフォーマビリティ
Guasto permanente	固定フォールト
Guasto fisico	物理フォールト
Manutenzione preventiva	予防保全
Prevenzione del guasto	（フォールト）予防
Trattamento dell'errore	誤り処理
Conseguimento della garanzia di funzionamento	デペンダビリティの達成方法
Test casuale	ランダム・テスト法
Funzione in tempo reale	実時間機能
Servizio in tempo reale	実時間での仕事
Sistema in tempo reale	実時間システム
Recupero dall'errore	（誤り）回復

English	French	German
Recovery point	Point de reprise	Rücksetzpunkt
Regression verification	Vérification de non-régression	Regressionsverifikation
Related faults	Fautes corrélées	Verwandte Fehler(ursachen)
Reliability	Fiabilité	Funktionsfähigkeit / Überlebenswahrscheinlichkeit
Reliability growth	Croissance de fiabilité	Überlebenswahrscheinlichkeit-wachstum
Removal (fault ~)	Elimination (des fautes)	Beseitigung (Fehler~)
Restoration (service ~)	Restauration (du service)	Wiederherstellung (Leistungs~)
Safety	Sécurité-innocuité	Sicherheit
Security	Sécurité-confidentialité	Vertraulichkeit / (Daten-Sicherheit)
Self-checking component	Composant auto-testable	Selbstprüfende Komponente
Sequential failures	Défaillances séquentielles	Folgeausfälle
Service	Service	Leistung
Severity (failure ~)	Sévérité (d'une défaillance)	Schweregrad (Ausfall~)
Simultaneous failures	Défaillances simultanées	Simultane Ausfälle
Soft fault	Faute douce	Weicher Fehler
Solid fault	Faute solide	Harter Fehler
Specification (system ~)	Spécification (d'un système)	Spezifikation (System~)
State (system ~)	Etat (d'un système)	Zustand (System~)
Stable reliability	Fiabilité stabilisée	Stabile Überlebenswahrscheinlichkeit
Static verification	Vérification statique	Statische Verifikation
Statistical testing	Test statistique	Statistisches Testen
Stopping failure	Défaillance par arrêt	Stillstand-Ausfall
Structure (system ~)	Structure (d'un système)	Struktur (System~)
Structural testing	Test structurel	Strukturelles Testen
Symbolic execution	Exécution symbolique	Symbolische Ausführung
System	Système	System
Temporary fault	Faute temporaire	Temporärer Fehler
Testing	Test	Testen
Timing failure	Défaillance temporelle	Zeitbezogener Ausfall / Zeit-Ausfall
Tolerance (fault ~)	Tolérance aux fautes	Toleranz (Fehler~)
Transient fault	Faute transitoire	Transienter Fehler
Treatment (fault ~)	Traitement (de faute)	Behandlung (Fehler~)
Trustability	Crédibilité	Vertrauenswürdigkeit
User (system ~)	Utilisateur (d'un système)	Benutzer (System~)
Validation	Validation	Validation
Value failure	Défaillance en valeur	Ergebnisbezogener Ausfall / Werte-Ausfall
Verification	Vérification	Verifikation

Italian	**Japanese**
Punto di recupero	回復点
Verifica di regressione	復帰検証
Guasti correlati	関連複数フォールト
Affidabilita'	信頼性、信頼度
Crescita di affidabilita'	信頼度成長
Eliminazione del guasto	(フォールト) 除去
Ripristino del servizio	(仕事) 復旧
Sicurezza di funzionamento	安全性、安全度
Sicurezza-confidenzialita'	セキュリティ
Componente che si autocontrolla	セルフチェッキング構成要素
Fallimenti sequenziali	逐次複数障害
Servizio	仕事、サービス
Gravita' del fallimento	(障害) 程度
Fallimenti simultanei	複数同時障害
Guasto debole	ソフトフォールト
Guasto solido	一定フォールト
Specifica del sistema	(システム) 仕様
Stato del sistema	(システム) 状態
Affidabilita' stabile	安定信頼度
Verifica statica	静的検証
Test statistico	統計的テスト法
Fallimento con blocco	停止障害
Struttura del sistema	(システム) 構造
Test strutturale	構造テスト法
Esecuzione simbolica	シンボリック実行
Sistema	システム
Guasto temporaneo	一時的フォールト
Test	テスト法
Fallimento nel tempo	タイミング障害
Tolleranza al guasto	(フォールト) トレランス
Guasto transitorio	過渡フォールト
Trattamento del guasto	(フォールト) 処置
Attendibilita'	信用性
Utente	(システム) 利用者
Validazione	確証 (確認)
Fallimento nel valore	値の障害
Verifica	検証

GPSR Compliance

The European Union's (EU) General Product Safety Regulation (GPSR) is a set of rules that requires consumer products to be safe and our obligations to ensure this.

If you have any concerns about our products, you can contact us on ProductSafety@springernature.com

In case Publisher is established outside the EU, the EU authorized representative is:

Springer Nature Customer Service Center GmbH
Europaplatz 3
69115 Heidelberg, Germany

Batch number: 10726298

Printed by Printforce, the Netherlands